T0236026

Automata Theory with Modern Applications

Recent applications to biomolecular science and DNA computing have created a new audience for automata theory and formal languages. This is the only introductory book to cover such applications. It begins with a clear and readily understood exposition of the basic principles, that assumes only a background in discrete mathematics. The first five chapters give a gentle but rigorous coverage of regular languages and Kleene's Theorem, minimal automata and syntactic monoids, Turing machines and decidability, and explain the relationship between context-free languages and pushdown automata. They include topics not found in other texts at this level, including codes, retracts, and semiretracts. The many examples and exercises help to develop the reader's insight. Chapter 6 introduces combinatorics on words and then uses it to describe a visually inspired approach to languages that is a fresh but accessible area of current research. The final chapter explains recently-developed language theory coming from developments in bioscience and DNA computing.

With over 350 exercises (for which solutions are available), plenty of examples and illustrations, this text will be welcomed by students as a contemporary introduction to this core subject; others, new to the field, will appreciate this account for self-learning.

Automata Theory
with Modern Applications

With contributions by Tom Head

JAMES A. ANDERSON

University of South Carolina Upstate

CAMBRIDGE
UNIVERSITY PRESS

University Printing House, Cambridge CB2 8BS, United Kingdom

One Liberty Plaza, 20th Floor, New York, NY 10006, USA

477 Williamstown Road, Port Melbourne, VIC 3207, Australia

314-321, 3rd Floor, Plot 3, Splendor Forum, Jasola District Centre, New Delhi - 110025, India

79 Anson Road, #06-04/06, Singapore 079906

Cambridge University Press is part of the University of Cambridge.

It furthers the University's mission by disseminating knowledge in the pursuit of education, learning and research at the highest international levels of excellence.

www.cambridge.org
Information on this title: www.cambridge.org/9780521613248

© Cambridge University Press 2006

First published 2006

A catalogue record for this publication is available from the British Library

ISBN 978-0-521-84887-9 Hardback
ISBN 978-0-521-61324-8 Paperback

Contents

Preface

This book serves two purposes, the first is as a text and the second is for someone wishing to explore topics not found in other automata theory texts. It was originally written as a text book for anyone seeking to learn the basic theories of automata, languages, and Turing machines. In the first five chapters, the book presents the necessary basic material for the study of these theories. Examples of topics included are: regular languages and Kleene's Theorem; minimal automata and syntactic monoids; the relationship between context-free languages and pushdown automata; and Turing machines and decidability. The exposition is gentle but rigorous, with many examples and exercises (teachers using the book with their course may obtain a copy of the solution manual by sending an email to solutions@cambridge.org). It includes topics not found in other texts such as codes, retracts, and semiretracts.

Thanks primarily to Tom Head, the book has been expanded so that it should be of interest to people in mathematics, computer science, biology, and possibly other areas. Thus, the second purpose of the book is to provide material for someone already familiar with the basic topics mentioned above, but seeking to explore topics not found in other automata theory books.

The two final chapters introduce two programs of research not previously included in beginning expositions. Chapter 6 introduces a visually inspired approach to languages allowed by the unique representation of each word as a power of a primitive word. The required elements of the theory of combinatorics on words are included in the exposition of this chapter. This is an entirely fresh area of research problems that are accessible on the completion of Chapter 6. Chapter 7 introduces recently developed language theory that has been inspired by developments in the biomolecular sciences and DNA computing. Both of these final chapters are kept within automata theory through their concentration on results in regular languages. Research in progress has begun to extend these concepts to broader classes of languages. There are now specialized books on

DNA-computing – and in fact a rapidly growing Springer-Verlag Series on 'Natural Computing' is in progress. This book is the first one to link (introductory) automata theory into this thriving new area.

Readers with a strong background will probably already be familiar with the material in Chapter 1. Those seeking to learn the basic theory of automata, languages, and Turing machines will probably want to read the chapters in order. The sections on retracts and semiretracts, while providing interesting examples of regular languages, are not necessary for reading the remainder of the book.

A person already familiar with the basics of automata, languages, and Turing machines, will probably go directly to Chapters 6 and 7 and possibly the sections on retracts and semiretracts.

I thank Tom Head for the work he has done on this book including his contributions of Chapters 6 and 7 as well as other topics. I also thank Brett Bernstein for his excellent proofreading of an early version of the book and Kristin and Phil Muzik for creating the figures for the book. Finally I would like to thank Ken Blake and David Tranah at Cambridge University Press for their help and support.

1
Introduction

1.1 Sets

Sets form the foundation for mathematics. We shall define a set to be a well-defined collection of objects. This definition is similar to the one given by Georg Cantor, one of the pioneeers in the early development of set theory. The inadequacy of this definition became apparent when paradoxes or contradictions were discovered by the Italian logician Burali-Forti in 1879 and later by Bertrand Russell with the famous Russell paradox. It became obvious that sets had to be defined more carefully. Axiomatic systems have been developed for set theory to correct the problems discussed above and hopefully to avoid further contradictions and paradoxes. These systems include the Zermelo–Fraenkel–von Neumann system, the Gödel–Hilbert–Bernays system and the Russell–Whitehead system. In these systems the items that were allowed to be sets were restricted. Axioms were created to define sets. Any object which could not be created from these axioms was not allowed to be a set. These systems have been shown to be equivalent in the sense that if one system is consistent, then they all are. However, Gödel has shown that if the systems are consistent, it is impossible to prove that they are.

Definition 1.1 *An object in a set is called an **element** of the set or is said to **belong** to the set. If an object x is an element of a set A, this is denoted by $x \in A$. If an object x is not a member of a set A, this is denoted by $x \notin A$.*

Objects in a set are called elements. Finite sets may be described by listing their elements. For example the set of positive integers less than or equal to seven may be described by the notation $\{1, 2, 3, 4, 5, 6, 7\}$ where the braces are used to indicate that we are describing a set. Thus symbols in an alphabet can be listed using this notation. We can also list the set of positive integers less than or equal to $10\,000$, by using the notation $\{1, 2, 3, 4, \ldots, 10\,000\}$ and the set of

1

positive integers by $\{1, 2, 3, 4, \ldots\}$, where three dots denote the continuation of a pattern. By definition, $1 \in \{1, 2, 3, 4, 5\}$ but $8 \notin \{1, 2, 3, 4, 5\}$. An element of a set may also be a set. Therefore $A = \{1, 2, \{3, 4, 5\}, 3, 4\}$ is a set that contains elements 1, 2, $\{3, 4, 5\}$, 3, and 4. Note that $5 \notin A$, but $\{3, 4, 5\} \in A$.

In many cases, listing the elements of a set can be tedious if not impossible. For example, consider listing the set of all primes. We thus have a second form of notation called **set builder notation**. Using this notation, the set of all objects having property P will be described by $\{x : x$ has property $P\}$. For example the set of all former Prime Ministers of Britain would by described by $\{x : x$ has been a Prime Minister of Britain$\}$. The set of all positive even integers less that or equal to 100, could be described by $\{x : x$ is a positive even integer less than or equal to 100$\}$.

Definition 1.2 *A set A is called a **subset** of a set B if every element of the set A is an element of the set B. If A is a subset of B, this is denoted by $A \subseteq B$. If A is not a subset of B, this is denoted by $A \nsubseteq B$.*

Therefore $\{a, b, c\} \subseteq \{a, b, c, d, e\}$ but $\{a, b, f\} \nsubseteq \{a, b, c, d, e\}$. By definition, any set is a subset of itself.

Definition 1.3 *A set A is **equal** to a set B if $A \subseteq B$ and $B \subseteq A$.*

Therefore two sets are equal if they contain the same elements. Notice that there is no order in a set. A set is simply defined by the elements that it contains. Also an element either belongs to a set or does not. It would be redundant to list an element more than once when defining a set.

Definition 1.4 *The intersection of two sets A and B, denoted by $A \cap B$, is the set consisting of all elements contained in both A and B.*

Let $A = \{x : x$ plays tennis$\}$ and $B = \{x : x$ plays golf$\}$, then $A \cap B = \{x : x$ plays tennis and golf$\}$. If $A = \{x : x$ is a positive integer divisible by 3$\}$ and $B = \{x : x$ is a positive integer divisible by 2$\}$, then $A \cap B = \{x : x$ is a positive integer divisible by 6$\}$.

Definition 1.5 *The union of two sets A and B, denoted by $A \cup B$, is the set consisting of all elements contained in either A or B.*

Let $A = \{x : x$ plays tennis$\}$ and $B = \{x : x$ plays golf$\}$, then $A \cup B = \{x : x$ plays tennis or golf$\}$.

If $A = \{x : x$ is a positive integer divisible by 3$\}$ and $B = \{x : x$ is a positive integer divisible by 2$\}$, then $A \cup B = \{x : x$ is a positive integer divisible by either 2 or 3$\}$.

Definition 1.6 *The **set difference**, denoted by $B - A$, is the set of all elements in the set B that are not in the set A.*

For example, the set $\{1, 2, 3, 4, 5\} - \{2, 4, 6, 8, 10\} = \{1, 3, 5\}$.

Example 1.1 Let $A = \{x : x$ plays tennis$\}$ and $B = \{x : x$ plays golf$\}$, the set $A - B = \{x : x$ plays tennis but does not play golf$\}$.

Definition 1.7 *The **symmetric difference**, denoted by $A \triangle B$, is the set $(A - B) \cup (B - A)$.*

It is easily seen that $A \triangle B = (A \cup B) - (A \cap B)$.

Example 1.2 Let $A = \{x : x$ plays tennis$\}$ and $B = \{x : x$ plays golf$\}$, the set $A \triangle B = \{x : x$ plays tennis or golf but not both$\}$.

We define two special sets. The first is the **empty set**, which is denoted by $\emptyset$ or $\{\}$. As the name implies, this set contains no elements. It is a subset of every set A since every element in the empty set is also in A. The second special set is the **universe** or **universe of discourse**, which we denote by $\mathcal{U}$. The universe is given, and limits or describes the type of sets under discussion, since they must all be subsets of the universe. For example if the sets we are describing are subsets of the integers then the universe could be the set of integers. If the universe is the the set of college students, then the set $\{x : x$ is a musician$\}$ would be the set of all musicians who are in college. Often the universe is understood and so is not explicitly mentioned. Later we shall see that the universe of particular interest to us is the set of all strings of symbols in a given alphabet.

Definition 1.8 *Let A be a set. $A' = \mathcal{U} - A$ is the set of all elements not in A.*

Example 1.3 Let A be the set of even integers and $\mathcal{U}$ be the set of integers. Then A' is the set of odd integers.

Example 1.4 Let $A = \{x : x$ collects coins$\}$, then $A' = \{x : x$ does not collect coins$\}$.

The proof of the following theorem is left to the reader.

Theorem 1.1 *Let A, B, and C be subsets of the universal set $\mathcal{U}$*

(a) Distributive properties

$$A \cap (B \cup C) = (A \cap B) \cup (A \cap C),$$
$$A \cup (B \cap C) = (A \cup B) \cap (A \cup C).$$

*(b) **Idempotent properties***

$$A \cap A = A,$$
$$A \cup A = A.$$

*(c) **Double Complement property***

$$(A')' = A.$$

*(d) **De Morgan's laws***

$$(A \cup B)' = A' \cap B',$$
$$(A \cap B)' = A' \cup B'.$$

*(e) **Commutative properties***

$$A \cap B = B \cap A,$$
$$A \cup B = B \cup A.$$

*(f) **Associative laws***

$$A \cap (B \cap C) = (A \cap B) \cap C,$$
$$A \cup (B \cup C) = (A \cup B) \cup C.$$

*(g) **Identity properties***

$$A \cup \emptyset = A,$$
$$A \cap \mathcal{U} = A.$$

*(h) **Complement properties***

$$A \cup A' = \mathcal{U},$$
$$A \cap A' = \emptyset.$$

Definition 1.9   *The **size** or **cardinality** of a finite set A, denoted by $|A|$, is the number of elements in the set. An infinite set which can be listed so that there is a first element, second element, third element etc. is called **countably infinite**. If it cannot be listed, it is said to be **uncountable**. Two infinite sets have the same **cardinality** if there is a one-to-one correspondence between the two sets. We denote this by $|A| = |B|$. If there is a one-to-one correspondence between A and a subset of B, we denote this by $|A| \leq |B|$. If $|A| \leq |B|$ but there is no one-to-one correspondence between A and B, then we denote this by $|A| < |B|$.*

Thus the cardinality of the set $\{a, b, c, \{d, e, f\}\}$ is 4. Intuitively, there is a one-to-one correspondence between two sets if elements of the two sets can be written in pairs so that each element in one set can be paired with one and only one element of the other set. The positive integers are obviously countable. Although it will not be proved here, the integers and rational numbers are both countable sets. The real numbers however are not a countable set. We see that there are two infinite sets, the countable sets and the uncountable sets with different cardinality; however, we shall soon see that there are an infinite number of infinite sets of different cardinality.

Further discussion of cardinality will be continued in the appendices.

Definition 1.10 *Let A and B be sets. The **Cartesian product** of A and B, denoted by $A \times B$ is the set $\{(a, b) : a \in A$ and $b \in B\}$.*

For example, let $A = \{a, b\}$ and $B = \{1, 2, 3\}$, then

$$A \times B = \{(a, 1)(a, 2)(a, 3)(b, 1)(b, 2)(b, 3)\}.$$

The familiar Cartesian plane $R \times R$ is the set of all ordered pairs of real numbers. Note that for finite sets $|A \times B| = |A| \times |B|$.

Definition 1.11 *The **power set** of a set A, denoted by $\mathcal{P}(A)$, is the set of all subsets of A.*

For example the power set of $\{a, b, c\}$ is

$$\{\{a\}, \{b\}, \{c\}, \{a, b\}, \{a, c\}, \{b, c\}, \{a, b, c\}, \emptyset\}.$$

In the finite case, it can be easily shown that $|\mathcal{P}(A)| = 2^{|A|}$.

Exercises

(1) State which of the following are true and which are false:
   (a) $\{\emptyset\} \subseteq A$ for an arbitrary set A.
   (b) $\emptyset \subseteq A$ for an arbitrary set A.
   (c) $\{a, b, c\} \subseteq \{a, b, \{a, b, c\}\}$.
   (d) $\{a, b, c\} \in \{a, b, \{a, b, c\}\}$.
   (e) $A \in \mathcal{P}(A)$.
(2) Prove Theorem 1.1. Let A, B, and C be subsets of the universal set $\mathcal{U}$.
   (a) **Idempotent property**
   $$A \cap A = A,$$
   $$A \cup A = A.$$

   (b) **Double Complement property**
   $$(A')' = A.$$

   (c) **De Morgan's laws**
   $$(A \cup B)' = A' \cap B',$$
   $$(A \cap B)' = A' \cup B'.$$

(d) **Commutative properties**

$$A \cap B = B \cap A,$$
$$A \cup B = B \cup A.$$

(e) **Associative properties**

$$A \cap (B \cap C) = (A \cap B) \cap C,$$
$$A \cup (B \cup C) = (A \cup B) \cup C.$$

(f) **Distributive properties**

$$A \cap (B \cup C) = (A \cap B) \cup (A \cap C),$$
$$A \cup (B \cap C) = (A \cup B) \cap (A \cup C).$$

(g) **Identity properties**

$$A \cup \emptyset = A,$$
$$A \cap \mathcal{U} = A.$$

(h) **Complement properties**

$$A \cup A' = \mathcal{U},$$
$$A \cap A' = \emptyset.$$

(3) Given a set $A \in \mathcal{P}(C)$, find a set B such that $A \triangle B = \emptyset$.

(4) If $A \subseteq B$, what is $A \triangle B$?

(5) Using the properties in Theorem 1.1 prove that $A \cap (B \triangle C) = (A \cap B) \triangle (A \cap C)$.

(6) Use induction to prove that for any finite set A, $|A| < |\mathcal{P}(A)|$.

(7) (Russell's Paradox) Let S be the set of all sets. Then $S \in S$. Obviously $\emptyset \notin \emptyset$. Let $W = \{A : A \notin A\}$. Discuss whether $W \in W$.

(8) Prove using the properties in Theorem 1.1
   (a) $A - (B \cup C) = (A - B) \cap (A - C)$,
   (b) $A - (B \cap C) = (A - B) \cup (A - C)$.

(9) Use the fact that $A \cap (A \cup B) = A$ to prove that $A \cup (A \cap B) = A$.

(10) Prove that if two disjoint sets are countable, then their union is countable.

1.2 Relations

Definition 1.12 *Given sets A and B, any subset $\mathcal{R}$ of $A \times B$ is a **relation** between A and B. If $(a, b) \in \mathcal{R}$, this is often denoted by $a\mathcal{R}b$. If $A = B$, $\mathcal{R}$ is said to be a **relation** on A.*

Note that relations need not have any particular property nor even be describable. Obviously we will be interested in those relations which are describable and have particular properties which will be shown later.

Example 1.5 If $A = \{a, b, c, d, e\}$ and $B = \{1, 2, 3, 4, 5\}$, then

$$\{(a, 3), (a, 2), (c, 2), (d, 4), (e, 4), (e, 5)\}$$

is a relation between A and B.

Example 1.6 $\{(x, y) : x \geq y\}$ and $\{(x, y) : x^2 + y^2 = 4\}$ are relations on R.

Example 1.7 If A is the set of people, then $a\mathcal{R}b$ if a and b are cousins is a relation on A.

Definition 1.13 *The **domain** of a relation $\mathcal{R}$ between A and B is the set $\{a : a \in A$ and there exists $b \in B$ so that $a\mathcal{R}b\}$. The **range** of a relation $\mathcal{R}$ between A and B is the set $\{b : b \in B$ and there exists $a \in A$ so that $a\mathcal{R}b\}$.*

Example 1.8 The domain and range of the relation $\{(x, y) : x^2 + y^2 = 4\}$ are $-2 \leq x \leq 2$ and $-2 \leq y \leq 2$ respectively.

Example 1.9 The relation $\mathcal{R}$ is on the set of people. The domain and range of $\mathcal{R}$ is the set of people who have cousins.

Definition 1.14 *Let $\mathcal{R}$ be a relation between A and B. The inverse of the relation $\mathcal{R}$ denoted by $\mathcal{R}^{-1}$ is a relation been B and A, defined by $\mathcal{R}^{-1} = \{(b, a) : (a, b) \in \mathcal{R}\}$.*

Example 1.10 If $A = \{a, b, c, d, e\}$ and $B = \{1, 2, 3, 4, 5\}$, and

$$\mathcal{R} = \{(a, 3), (a, 2), (b, 3), (b, 5), (c, 3), (d, 2), (d, 3), (e, 4), (e, 5)\}$$

is a relation between A and B then

$$\mathcal{R}^{-1} = \{(3, a), (2, a), (3, b), (5, b), (3, c), (2, d), (3, d), (4, e), (5, e)\}$$

is a relation between B and A.

Example 1.11 If $\mathcal{R} = \{(x, y) : y = 4x^2\}$, then $\mathcal{R}^{-1} = \{(y, x) : y = 4x^2\}$.

Definition 1.15 *Let $\mathcal{R}$ be a relation between A and B, and let $\mathcal{S}$ be a relation between B and C. The composition of $\mathcal{R}$ and $\mathcal{S}$, denoted by $\mathcal{S} \circ \mathcal{R}$ is a relation between A and C defined by $(a, c) \in \mathcal{S} \circ \mathcal{R}$ if there exists $b \in B$ such that $(a, b) \in \mathcal{R}$ and $(b, c) \in \mathcal{S}$.* •

Example 1.12 Let $A = \{a, b, c, d, e\}$ and $B = \{1, 2, 3, 4, 5\}$ and

$$\mathcal{R} = \{(a, 3), (a, 2), (c, 2), (d, 4), (e, 4), (e, 5)\}$$

be a relation between A and B. Then, as shown above

$$\mathcal{R}^{-1} = \{(3, a), (2, a), (2, c), (4, d), (4, e), (5, e)\}$$

is a relation between B, and A,

$$\mathcal{R} \circ \mathcal{R}^{-1} = \{(3, 3), (3, 2), (2, 2), (2, 3), (4, 4), (5, 5)\}$$

is a relation on B, and

$$\mathcal{R}^{-1} \circ \mathcal{R} = \{(a, a), (a, c), (c, a), (c, c), (d, d), (d, e), (e, e)\}$$

is a relation on A.

Example 1.13　If $\mathcal{R} = \{(x, y) : y = x + 5\}$ and $\mathcal{S} = \{(y, z) : z = y^2\}$ then $\mathcal{S} \circ \mathcal{R} = \{(x, z) : z = (x + 5)^2\}$.

Theorem 1.2　*Composition of relations is associative; that is, if A, B, and C are sets and if $R \subseteq A \times B$, $S \subseteq B \times C$, and $T \subseteq C \times D$, then $T \circ (S \circ R) = (T \circ S) \circ R$.*

Proof　First show that $T \circ (S \circ R) \subseteq (T \circ S) \circ R$. Let $(a, d) \in T \circ (S \circ R)$, then there exists $c \in C$ such that $(a, c) \in S \circ R$ and $(c, d) \in T$. Since $(a, c) \in S \circ R$, there exists $b \in B$ so that $(a, b) \in R$ and $(b, c) \in S$. Since $(b, c) \in S$ and $(c, d) \in T$, $(b, d) \in T \circ S$. Since $(b, d) \in T \circ S$ and $(a, b) \in R$, $(a, d) \in (T \circ S) \circ R$. Thus, $T \circ (S \circ R) \subseteq (T \circ S) \circ R$. The second part of the proof showing that $(T \circ S) \circ R \subseteq T \circ (S \circ R)$ is similar and is left to the reader.　□

When $\mathcal{R}$ is a relation on a set A, there are certain special properties that $\mathcal{R}$ may have which we now consider.

Definition 1.16　*A relation $\mathcal{R}$ on A is **reflexive** if $a\mathcal{R}a$ for all $a \in A$. A relation $\mathcal{R}$ on A is **symmetric** if $a\mathcal{R}b \to b\mathcal{R}a$ for all $a, b \in A$. A relation $\mathcal{R}$ on A is **antisymmetric** if $a\mathcal{R}b$ and $b\mathcal{R}a$ implies $a = b$. A relation is **transitive** if whenever $a\mathcal{R}b$ and $b\mathcal{R}c$, then $a\mathcal{R}c$.*

Example 1.14　Let A be the set of all people and $a\mathcal{R}b$ if a and b are siblings. The relation $\mathcal{R}$ is not reflexive since a person cannot be their own brother or sister. It is symmetric however since if a and b are siblings, then b and a are siblings. It might appear that $\mathcal{R}$ is transitive. Such is not the case however since if a and b are siblings, and b and a are siblings, we must conclude that a and a are siblings, which we know is not true.

Example 1.15　Let A be the set of all people and $a\mathcal{R}b$ if a and b have the same parents. The relation $\mathcal{R}$ is reflexive since everyone has the same parents as themselves. It is symmetric since if a and b have the same parents, b and *a* have the same parents. It is also transitive since if *a* and *b* have the same parents and *b* and *c* have the same parents, then *a* and *c* have the same parents.

Example 1.16   Let $A = \{a, b, c, d, e\}$ and

$$\mathcal{R} = \{(a, a), (a, b), (b, c), (b, b), (a, c), (c, c), (d, d), (a, d), (c, e), (d, a), (b, a)\}.$$

$\mathcal{R}$ is not reflexive since $(e, e) \notin \mathcal{R}$. It is not symmetric because $(a, c) \in \mathcal{R}$, but $(c, a) \notin \mathcal{R}$. It is not antisymmetric since $(a, d), (d, a) \in \mathcal{R}$, but $d \neq a$. It is not transitive since $(a, c), (c, e) \in \mathcal{R}$, but $(a, e) \notin \mathcal{R}$.

Example 1.17   Let $\mathcal{R}$ be the relation on Z defined by $a\mathcal{R}b$ if $a - b$ is a multiple of 5. Certainly $a - a = 0$ is a multiple of 5, so $\mathcal{R}$ is reflexive. If $a - b$ is a multiple of 5, then $a - b = 5k$ for some integer k. Hence $b - a = 5(-k)$ is a multiple of 5, so $\mathcal{R}$ is symmetric. If $a - b$ is a multiple of 5 and $b - c$ is a multiple of 5, then $a - b = 5k$ and $b - c = 5m$ for some integers k and m.

$$a - c = a - b + b - c$$
$$= 5k + 5m$$
$$= 5(k + m)$$

so that $a - c$ is a multiple of 5. Hence $\mathcal{R}$ is transitive.

Definition 1.17   *A relation $\mathcal{R}$ on A is an **equivalence relation** if it is reflexive, symmetric, and transitive.*

Example 1.18   Let Z be the set of integers and $\mathcal{R}_1$ be the relation on Z defined by $\mathcal{R}_1 = \{(m, n) : m - n\}$ is divisible by 5. $\mathcal{R}_1$ is shown above to be an equivalence relation on the integers.

Example 1.19   Let A be the set of all people. Define $\mathcal{R}_2$ by $a\mathcal{R}_2b$ if a and b are the same age. This is easily shown to be an equivalence relation.

An equivalence relation on a set A divides A into nonempty subsets that are **mutually exclusive** or **disjoint**, meaning that no two of them have an element in common. In the first example above, the sets

$$\{\ldots - 20, -15, -10, -5, 0, 5, 10, 15, 20, \ldots\}$$
$$\{\ldots - 19, -14, -9, -4, 1, 6, 11, 16, 21, \ldots\}$$
$$\{\ldots - 18, -13, -8, -3, 2, 7, 12, 17, 22, \ldots\}$$
$$\{\ldots - 17, -12, -7, -2, 3, 8, 13, 18, 23, \ldots\}$$
$$\{\ldots - 18, -11, -6, -1, 4, 9, 14, 19, 24, \ldots\}$$

contain elements that are related to each other and no element in one set is related to an element in another set. In the second example the sets $\{s_n = x : x$ is n years old$\}$ for $n = 0, 1, 2, \ldots$ also divide the set of people into sets that are related to each other. Also no person can belong to two sets. (See the definition of partition below.)

Notation 1.1   Let R be an equivalence relation on a set A and $a \in A$. Then $[a]_\mathcal{R} = \{x : x\mathcal{R}a\}$. If the relation is understood, then $[a]_\mathcal{R}$ is simply denoted by $[a]$. Let $[A]_\mathcal{R} = \{[a]_\mathcal{R} : a \in A\}$.

Definition 1.18   *Let A and I be nonempty sets and $\langle A \rangle = \{A_i : i \in I\}$ be a set of nonempty subsets of A. The set $\langle A \rangle$ is called a **partition** of A if both of the following are satisfied:*

(a) $A_i \cap A_j = \emptyset$ for all $i \neq j$.
(b) $A = \bigcup_{i \in I} A_i$; that is, $a \in A$ if and only if $a \in A_i$ for some $i \in I$.

Theorem 1.3   *A nonempty set of subsets $\langle A \rangle$ of a set A is a partition of A if and only if $\langle A \rangle = [A]_\mathcal{R}$ for some equivalence relation $\mathcal{R}$.*

Proof   Let $\langle A \rangle = \{A_i : i \in I\}$ be a partition of A. Define a relation $\mathcal{R}$ on A by $a\mathcal{R}b$ if and only if a and b are in the same subset A_i for some i. Certainly for all a in A, $a\mathcal{R}a$ and $\mathcal{R}$ is reflexive. If a and b are in the same subset A_i, then b and a are in the subset A_i and $\mathcal{R}$ is symmetric. Since the sets $A_i \cap A_j = \emptyset$ for $i \neq j$, if a and b are in the same subset and b and c are in the same subset, then a and c are in the same subset. Hence $\mathcal{R}$ is transitive and $\mathcal{R}$ is an equivalence relation.

Conversely, assume that $\mathcal{R}$ is an equivalence relation. We need to show that $[A]_\mathcal{R} = \{[a] : a \in A\}$ is a partition of A. Certainly, for all a, $[a]$ is nonempty since $a \in [a]$. Obviously, A is the union of the $[a]$, such that $a \in A$. Assume that $[a] \cap [b]$ is nonempty and let $x \in [a] \cap [b]$. Then $x\mathcal{R}a$ and $x\mathcal{R}b$, and by symmetry, $a\mathcal{R}x$. But since $a\mathcal{R}x$ and $x\mathcal{R}b$, by transitivity, $a\mathcal{R}b$. Therefore, $a \in [b]$. If $y \in [a]$, then $y\mathcal{R}a$ and since $a\mathcal{R}b$, by transitivity, $y\mathcal{R}b$. Therefore, $[a] \subseteq [b]$. Similarly, $[b] \subseteq [a]$ so that $[a] = [b]$, and we have a partition of A. $\square$

Definition 1.19   *$[A]_\mathcal{R}$ is called the set of **equivalence classes** of A given by the relation $\mathcal{R}$.*

If the symmetric property is changed to antisymmetric property, we have the following:

Definition 1.20   *A relation $\mathcal{R}$ on A is a **partial ordering** if it is reflexive, antisymmetric, and transitive. If $\mathcal{R}$ is a partial ordering on A, then $(A, \mathcal{R})$ is called a **partially ordered set** or a **poset**.*

Example 1.20 Let A be collection of subsets of a set S. Define the relation $\leq$ by $U \leq V$ if $U \subseteq V$. It is easily seen that $(A, \subseteq)$ is a partially ordered set.

Example 1.21 Let R be the set of real numbers. Define the relation $\leq$ by $r \leq s$ if r is less than or equal to s using the usual ordering on R.

Definition 1.21 *Let $(A, \leq)$ be a partially ordered set. If $a, b \in A$ and either $a \leq b$ or $b \leq a$ then a and b are said to be comparable. If for every $a, b \in A$, a and b are comparable then $(A, \leq)$ is called a **chain** or a **total ordering**.*

Definition 1.22 *For a subset B of a poset A, an element a of A is an **upper bound** of B if $b \leq a$ (or $a \geq b$) for all b in B. The element a is called a **least upper bound** (lub) of B if (i) a is an upper bound of B and (ii) if any other element a' of A is an upper bound of B, then $a \leq a'$. The least upper bound for the entire poset A (if it exists) is called the **greatest element** of A. For a subset B of a poset A, an element a of A is a **lower bound** of B if $a \leq b$ (or $b \geq a$) for all b in B. The element a is called a **greatest lower bound** (glb) of B if (i) a is a lower bound of B and (ii) if any other element a' of A is a lower bound of B, then $a \geq a'$. The greatest lower bound for the entire poset A (if it exists) is called the **least element** of A.*

Example 1.22 Let $C = \{a, b, c\}$ and X be the power set of C.

$$X = P(C) = \{\varnothing, \{a\}, \{b\}, \{c\}, \{a, b\}, \{a, c\}, \{b, c\}, \{a, b, c\}\}.$$

Define the relation $\leq$ on X by $T \leq V$ if $T \subseteq V$. By definition, $\{a, b\}$ is the greatest lower bound of $\{\varnothing, \{a\}, \{b\}\}$ and also of $\{\varnothing, \{a\}, \{b\}, \{a, b\}\}$. The set $\{a, b, c\}$ is the least upper bound of X. The element $\varnothing$ is the greatest lower bound for all three sets.

Definition 1.23 *A poset A for which every pair of elements of A have a least upper bound in A is called an **upper semilattice** and is denoted by $(\mathbf{A}, \vee)$ or $(A, +)$.*

If every two elements of a poset A have a greatest lower bound in A, then the following binary relation can be defined on the set. If a and b belong to A, let $a \wedge b = \mathrm{glb}\{a, b\}$.

Definition 1.24 *A poset A for which every pair of elements of A have a greatest lower bound in A is called a **lower semilattice** and is denoted by $(\mathbf{A}, \wedge)$ or $(A, \cdot)$.*

Example 1.22 is an example of a poset which is both an upper semilattice and a lower semilattice.

Exercises

(1) What is wrong with the following proof?

   If a relation R on a set A is symmetric and transitive, then it is reflexive.

   Proof   Since R is symmetric, if aRb then bRa. Since A is transitive, if aRb and bRa then aRa. Therefore aRa and R is reflexive.

(2) Give an example of a relation R on a set A that is reflexive and symmetric, but not transitive.

(3) Give an example of a relation R on a set A that is reflexive and transitive, but not symmetric.

(4) Let σ and τ be relations of a set A. Show that $\sigma \subseteq \tau$ if and only if each equivalence class in the set of equivalence classes given by τ is a union of equivalence classes given by σ.

(5) A relation R of A is a **partial order** if it is reflexive, antisymmetric, and transitive. It is a **total order** or **chain** if for any two elements $a, b \in A$, either aRb or bRa. Give an example of a partial order that is not a total order.

(6) Prove that the intersection of two partial orders on a set A is a partial order.

(7) Prove that the intersection of two equivalence relations on a set A is an equivalence relation.

(8) Given a set A, what is the intersection of all equivalence relations on A?

(9) Let A be the set of ordered pairs of positive integers. Define the relation R on A by $(a, b)R(c, d)$ if $ad = bc$. Is R an equivalence relation? If so what are the equivalence classes?

1.3   Functions

Definition 1.25   *A relation f on $A \times B$ is a **function** from A to B, denoted by $f : A \to B$, if for every $a \in A$ there is one and only one $b \in B$ so that $(a, b) \in f$. If $f : A \to B$ is a function and $(a, b) \in f$, we say that $b = f(a)$. The set A is called the **domain** of the function f and B is called the **codomain**. If $E \subseteq A$, then $f(E) = \{b : f(a) = b \text{ for some } a \text{ in } E\}$ is called the **image** of E. The image of A itself is called the **range** of f. If $F \subseteq B$, then $f^{-1}(F) = \{a : f(a) \in F\}$ is called the **preimage** of F. A function $f : A \to B$ is also called a **mapping** and we speak of the domain A being mapped into B by the mapping f. If $(a, b) \in f$ so that $b = f(a)$, then we say that the element a is mapped to the element b.*

The proof of the following theorem is left to the reader.

Theorem 1.4   *Let $f : A \to B$.*

(a) $f(A_1 \cup A_2) = f(A_1) \cup f(A_2)$ for $A_1, A_2 \subseteq A$.
(b) $f^{-1}(B_1 \cup B_2) = f^{-1}(B_1) \cup f^{-1}(B_2)$ for $B_1, B_2 \subseteq B$.
(c) $f(A_1 \cap A_2) \subseteq f(A_1) \cap f(A_2)$ for $A_1, A_2 \subseteq A$.
(d) $f^{-1}(B_1 \cap B_2) = f^{-1}(B_1) \cap f^{-1}(B_2)$ for $B_1, B_2 \subseteq B$.
(e) $f^{-1}(B_1') = (f^{-1}(B_1))'$ for $B_1 \subseteq B$.

Definition 1.26   *If $f : A \to B$, and the image of f is B, then f is **onto**. It is also called a **surjection** or an **epimorphism**. Thus for element $b \in B$, there is an element $a \in A$ so that $b = f(a)$.*

Definition 1.27   *If $f : A \to B$, and $f(a) = f(a') \Rightarrow a = a'$ for all $a, a' \in A$ then f is **one-to-one**. It is also called a **monomorphism** or **injection**.*

Definition 1.28   *If $f : A \to B$ is one-to-one and onto, then f is called a **one-to-one correspondence** or **bijection**. If A is finite, then f is also called a **permutation**.*

Notation 1.2   If f is a permutation on the set $\{1, 2, 3, \ldots, n\}$, then it can be represented in the form

$$\begin{pmatrix} 1 & 2 & \ldots & n \\ f(1) & f(2) & \ldots & f(n) \end{pmatrix}.$$

Thus if $f(a) = b$, $f(b) = d$, $f(c) = a$, and $f(d) = c$, we may denote this by $\begin{pmatrix} a & b & c & d \\ b & d & a & c \end{pmatrix}$. If $f = \begin{pmatrix} 1 & 2 & 3 & 4 \\ 3 & 4 & 1 & 2 \end{pmatrix}$ and $g = \begin{pmatrix} 1 & 2 & 3 & 4 \\ 2 & 3 & 4 & 1 \end{pmatrix}$, to find the composition $f \circ g$ note that since $g(1) = 2$ appears under 1 in the permutation for g, and $f(2) = 4$, appears under 2 in f, we may find $(f \circ g)(1)$ by going from 1 down to 2 in g and then going from 2 down to 4 in the permutation f, so $(f \circ g)(1) = 4$. Similarly, to find $(f \circ g)(2)$, go down from 2 to 3 in g and from 3 to 1 in f, so $(f \circ g)(2) = 1$. Continuing, we have $f \circ g = \begin{pmatrix} 1 & 2 & 3 & 4 \\ 4 & 1 & 2 & 3 \end{pmatrix}$.

Example 1.23   Let $f : A \to B$, where A and B are the set of real numbers, be defined by $f(x) = x^2$, then f is a function whose range is the set of nonnegative real numbers. It is not onto since the range is not B. It is not one-to-one since $f(2) = f(-2) = 4$.

Example 1.24   Let $f : A \to B$, where A and B are the set of real numbers, be defined by $f(x) = x^3$, then f is a function whose range is B. Hence it is onto. It is also one-to-one since $a^3 = (a')^3 \Rightarrow a = a'$.

Definition 1.29 *Let $I : A \to A$ be defined by $I(a) = a$ for all $a \in A$. The function I is called the **identity function**.*

Definition 1.30 *Let $g : A \to B$ and $f : B \to C$, then $(f \circ g)(x) = f(g(x))$.*

The proof of the following theorem is elementary and is left to the reader:

Theorem 1.5 *Let $f : A \to A$, then $I \circ f = f \circ I = f$.*

Theorem 1.6 *Let $f : A \to B$ then there exists a function $f^{-1} : B \to A$ so that $f \circ f^{-1} = f^{-1} \circ f = I$ if and only if f is a bijection. The function f^{-1} is also a bijection.*

Proof Assume there exists a function $f^{-1} : B \to A$ so that $f \circ f^{-1} = f^{-1} \circ f = I$, and $f(a) = f(a')$. Then $f^{-1} \circ f(a) = f^{-1} \circ f(a')$, so $I(a) = I(a')$ and $a = a'$. Therefore f is one-to-one. Let $b \in B$ and $a = f^{-1}(b)$. Then $f(a) = f(f^{-1}(b)) = b$, f is onto.

Assume $f : A \to B$ is a bijection. Define the relation f^{-1} on $B \times A$ by $f^{-1}(b) = a$ if $f(a) = b$. Let $b \in B$ and choose a so that $f(a) = b$. This is possible since f is onto. Therefore $f^{-1}(b) = a$ and f^{-1} has domain B. If $f^{-1}(b) = a$ and $f^{-1}(b) = a'$, then $f(a) = b$ and $f(a') = b$. But since f is one-to-one, $a = a'$. Therefore f^{-1} is well defined and hence f^{-1} is a function. By definition $f \circ f^{-1} = f^{-1} \circ f = I$.

By symmetry, f^{-1} is a bijection. □

The proof of the following theorem is left to the reader:

Theorem 1.7 *Let $g : A \to B$ and $f : B \to C$; then:*

(a) *If g and f are onto B and C, respectively, then $f \circ g$ is onto C.*
(b) *If g and f are both one-to-one, then $f \circ g$ is one-to-one.*
(c) *If g and f are both one-to-one and onto, then $f \circ g$ is one-to-one and onto.*
(d) *If g and f have inverses, then $(f \circ g)^{-1} = g^{-1} \circ f^{-1}$.*

Theorem 1.8 *Let $f : A \to B$ be a function. The relation R defined by aRa' if $f(a) = f(a')$ is an equivalence relation.*

Proof Let $a, a', a'' \in A$. Certainly $f(a) = f(a)$ so R is reflexive. If $f(a) = f(a')$, then $f(a') = f(a)$, so R is symmetric. If $f(a) = f(a')$ and $f(a') = f(a'')$, then $f(a) = f(a'')$ so R is transitive. Therefore R is an equivalence relation. □

Definition 1.31 *Let $\mathcal{R}$ be an equivalence relation on A, and $\phi_\mathcal{R} : A \to [A]_\mathcal{R}$ be a function defined by $\phi_\mathcal{R}(a) = [a]$. The function $\phi_\mathcal{R}$ is called the canonical function from A to $[A]_\mathcal{R}$.*

Theorem 1.9 *Let $f : A \to B$ be a function and $\mathcal{R}$ be the relation $a\mathcal{R}a'$ iff $f(a) = f(a')$, then there exists a function $g : [A]_{\mathcal{R}} \to B$ defined by $g(a_{\mathcal{R}}) = f(a)$. Hence $g \circ \phi_{\mathcal{R}} = f$.*

Proof Assume $g([a]) = b$ and $g([a]) = b'$, then $f(a) = b$ and $f(a') = b'$, where $[a] = [a']$. But then $a\mathcal{R}a'$ and $f(a) = f(a')$. Therefore $b = b'$ and g is a function. $\qquad\square$

Theorem 1.10 *Let $f : [A]_{\mathcal{R}} \to B$ be a function and S be an equivalence relation such that $S \subseteq \mathcal{R}$ and $aSa' \to a\mathcal{R}a'$, then there exist functions $g : [A]_S \to B$ and $i : [A]_S \to [A]_{\mathcal{R}}$ such that $f \circ i = g$.*

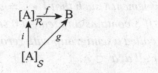

Proof Let $i : [A]_S \to [A]_{\mathcal{R}}$ be defined by $i([a]_S) = [a]_{\mathcal{R}}$ and $g : [A]_S \to B$ by $g([a]_S) = f([a_{\mathcal{R}}])$. The function i is trivially well defined. The proof that g is a function is similar to the proof of the previous theorem. $\qquad\square$

Exercises

(1) Prove Theorem 1.4. Let $f : A \to B$.
   (a) $f(A_1 \cup A_2) = f(A_1) \cup f(A_2)$ for $A_1, A_2 \subseteq A$.
   (b) $f^{-1}(B_1 \cup B_2) = f^{-1}(B_1) \cup f^{-1}(B_2)$ for $B_1, B_2 \subseteq B$.
   (c) $f(A_1 \cap A_2) \subseteq f(A_1) \cap f(A_2)$ for $A_1, A_2 \subseteq A$.
   (d) $f^{-1}(B_1 \cap B_2) = f^{-1}(B_1) \cap f^{-1}(B_2)$ for $B_1, B_2 \subseteq B$.
   (e) $f^{-1}(B_1') = (f^{-1}(B_1))'$ for $B_1 \subseteq B$.

(2) Prove Theorem 1.7. Let $g : A \to B$ and $f : B \to C$; then:
   (a) If g and f are onto B and C, respectively, then $f \circ g$ is onto C.
   (b) If g and f are both one-to-one, then $f \circ g$ is one-to-one.
   (c) If g and f are both one-to-one and onto, then $f \circ g$ is one-to-one and onto.
   (d) If g and f have inverses, then $(f \circ g)^{-1} = g^{-1} \circ f^{-1}$.

(3) Give an example of a function f and sets $A_1, A_2 \subseteq A$ such that $f(A_1 \cap A_2) \neq f(A_1) \cap f(A_2)$.

(4) Prove that if $f \circ g$ is one-to-one then g is one-to-one.

(5) Prove that if $f \circ g$ is onto, then f is onto.

1.4   Semigroups

In the following function $\star : S \times S \to S$ we shall use the notation $a \star a'$ for $\star((a, a'))$.

Definition 1.32   *A **semigroup** is a nonempty set S together with a function $\star$ from $S \times S \to S$ such that*

$$a \star (a' \star a'') = (a \star a') \star a''.$$

*The function or **operation** $\star$ with this property is called **associative**. The semigroup is denoted by $(S, \star)$ or simply S if the operation is understood. If S contains an identity element 1 such that $1 \star a = a \star 1 = a$ for all $a \in A$, then S is called a **monoid**. If S contains an element 0 such that $0 \star a = a \star 0 = 0$ for all $a \in A$, then S is called a **semigroup with zero**. A semigroup is **commutative** if $a \star a' = a' \star a$ for all $a, a' \in A$.*

Example 1.25   Examples of semigroups include

(1) The set of integers [positive integers, real numbers, positive real numbers, rational numbers, positive rational numbers] together with either of the operations addition or multiplication is a semigroup.

(2) The set of functions $\{ f \mid f : A \to A \}$ for a given set A, together with the operation $\circ$ where $(f \circ g)(x) = f(g(x))$ is a semigroup.

(3) The set of $n \times n$ matrices with either of the operations addition or multiplication is a semigroup.

Example 1.26   The set of nonnegative integers together with the operation addition is a monoid. All of the above examples except for the positive real numbers, positive integers, and positive rational numbers with the operation addition form a monoid.

Every semigroup S can be changed to a monoid by simply adding an element 1 and defining $1 \star a = a \star 1 = a$ for all $a \in S$. If S was already a monoid, it remains a monoid but with a different identity element. Normally one adds an identity element to a semigroup only if it is not already a monoid. Similarly every semigroup S can be changed to a semigroup with zero by simply adding an element 0 and define $0 \star a = a \star 0 = 0$ for all $a \in S$. Note that if we let S_m

be the set of all integers greater than or equal to m, then $(S_m, +)$ is a semigroup. If we include 0, then we have a monoid. Also since $(S_m, \cdot)$ is a semigroup, if we include 1, then we have a monoid.

Notation 1.3 *Let S be a semigroup. The monoid $S^1 = S$ if S is already a monoid, and $S^1 = S \cup \{1\}$ otherwise. The semigroup $S^0 = S$ if S is already a semigroup with zero and $S^0 = S \cup \{0\}$ otherwise.*

Definition 1.33 *Let $(S, \star)$ be a semigroup and H be a nonempty subset of S. If for all $h, h' \in H$, $h \star h' \in H$, then $(H, \star)$ is a **subsemigroup** of $(S, \star)$. If $(S, \star)$ is a monoid and $(H, \star)$ is a subsemigroup of $(S, \star)$ containing the identity of the monoid, then $(H, \star)$ is a **submonoid** of $(S, \star)$.*

Therefore the set of positive integers with the operation multiplication is a submonoid of the integers with the operation multiplication. The semigroup $(S_m, +)$ is a subsemigroup of $(S_n, +)$ for $m \leq n$.

Theorem 1.11 *Let $(S, \star)$ be a semigroup and $\{H_i : i \in I\}$ be subsemigroups of S. If the intersection $\bigcap_{i \in I} H_i$ is nonempty, then it is a subsemigroup of S.*

Proof Let $h, h' \in \bigcap_{i \in I} H_i$. Then $h, h' \in H_i$ for each $i \in I$ and $h \star h' \in H_i$ for each i. Therefore $h \star h' \in \bigcap_{i \in I} H_i$, and $\bigcap_{i \in I} H_i$ is a subsemigroup of S. $\qquad \square$

Corollary 1.1 *Let $(S, \star)$ be a monoid and $\{H_i : i \in I\}$ be submonoids of I. The intersection $\bigcap_{i \in I} H_i$ is a submonoid.*

Theorem 1.12 *Let $(S, \star)$ be a semigroup and W be a nonempty subset of S. There exists a smallest subsemigroup of S containing W.*

Proof Let H be the intersection of all subsemigroups of S containing W. By the previous theorem H is a subsemigroup and is contained in all other subsemigroups of S containing W. $\qquad \square$

Definition 1.34 *The smallest subsemigroup of S containing the nonempty set W is the semigroup **generated** by W. It is denoted by $\langle W \rangle$.*

The proof of the following theorem is left to the reader.

Theorem 1.13 *Let $(S, \star)$ be a semigroup and W be a nonempty subset of S. The set of all finite products of elements of W together with the elements of W is the semigroup generated by W.*

Definition 1.35 *Let $(M, \star)$ be a monoid and W be a nonempty subset of M. The semigroup generated by W, together with the identity of M is called the **monoid generated** by W. It is denoted by W^*.*

Definition 1.36 *A commutative semigroup* $(S, *)$ *is a **semilattice** if* $a * a = a$
for all $a \in S$. *An element a of a semigroup is called an **idempotent** element if*
$a * a = a$. *A semilattice is therefore a commutative semigroup in which every*
element is an idempotent. If $(S, *)$ *is a semilattice and* $\overline{S} \subseteq S$ *then* $\overline{S}$ *is a*
subsemilattice of S if $*$ *is a binary operation on* $\overline{S}$. *Equivalently,* $(\overline{S}, *)$ *is a*
subsemilattice of $(S, *)$ *if* $\overline{S} \subseteq S$ *and for every* $a, b \in \overline{S}$, $a * b \in \overline{S}$.

Example 1.27 The semigroup consisting of all subsets of a fixed set T
together with the operation $\cap$ is a semilattice.

Obviously lower semilattices and upper semilattices are semilattices. Conversely given a semilattice $(S, *)$, a partial ordering on S can be defined by
$s \leq t$ if $s * t = t$.

Definition 1.37 *If* $(S, *)$ *is both a lower semilattice and an upper semilattice*
*then it is called a **lattice**. If for any lattice* $(S, *)$ *and any subset* T *of S, both*
the greatest lower bound and the least upper bound exist, then $(S, *)$ *is called*
*a **complete lattice**.*

Definition 1.38 *A **group** G is a monoid such that for every* $g \in G$, *there exists*
$g^{-1} \in G$ *such that* $gg^{-1} = g^{-1}g = 1$ *where* 1 *is the identity of the monoid.*

If a semigroup $(S, \star)$ is infinite, then it is possible that the semigroup
generated by $\{a\}$ is infinite. It consists of the elements $\{a, a^2, a^3, \ldots\}$ where
$a^{k+1} = a^k \star a$. For example, if Z is the semigroup of integers under addition,
then the semigroup generated by $\{2\}$ is $\{2, 4, 6, 8, \ldots\}$. If a semigroup $(S, \star)$ is
finite, however, for some k and m, $a^k = a^{k+m}$. Pick the smallest k and m, then
$a^k, a^{k+1}, a^{k+2}, \ldots, a^{m-1}$ form a semigroup. If each element is multiplied by a^k
we again get $a^k, a^{k+1}, a^{k+2}, \ldots, a^{m-1}$ so there is some a^i so that $a^k \star a^i = a^k$.
Therefore $a^i \star a^{k+j} = a^{k+j}$ for all $0 \leq j \leq m - 1$ and a^i is the identity of the
semigroup. Therefore the semigroup is a monoid. Also for each a^j there exist
a^n such that $a^j \star a^n = a^n \star a^j = a^i$. This element is called the inverse of a^j.
Hence this set forms a group.

Definition 1.39 *A function f from the semigroup* $(S, \star)$ *to the semigroup* $(T, \bullet)$
*is called a **semigroup homomorphism** if* $f(s \star s') = f(s) \bullet f(s')$ *for all* $s, s' \in$
S. *If the semigroup f is one-to-one and onto, then f is called a **semigroup**
isomorphism. A function f from the monoid* $(S, \star)$ *to the monoid* $(T, \bullet)$ *is*
*called a **monoid homomorphism** if* $f(s \star s') = f(s) \bullet f(s')$ *for all* $s, s' \in S$
and f maps the identity of S to the identity of T. If a monoid homomorphism
*f is one-to-one and onto, then f is called a **monoid isomorphism**.*

Normally when a function is a homomorphism from a monoid to a monoid, we shall assume that it is a monoid homomorphism and simply call it a homomorphism.

Example 1.28   Let $f : (Z, +) \rightarrow (Z, +)$ be defined by $f(a) = 2a$, then f is a semigroup homomorphism. It is also a monoid homomorphism.

Example 1.29   Let $2Z$ be the set of even integers and $f : (Z, +) \rightarrow (2Z, +)$ be defined by $f(a) = 2a$, then f is a monoid homomorphism. It is also a monoid isomorphism.

Example 1.30   Let S be the semigroup of $n \times n$ matrices with the operation multiplication, R be the semigroup of real numbers with the operation multiplication, and $\det(A)$ be the determinant of a matrix A. Then $\det : (S, \cdot) \rightarrow (R, \cdot)$ is a homomorphism.

Example 1.31   Let R_+ denote the semigroup of positive real numbers with the operation multiplication and $\ln$ be the natural logarithm, then $\ln : (R_+, \cdot) \rightarrow (R, +)$ is a homomorphism.

The following theorem is left to the reader:

Theorem 1.14   *Let $f : S \rightarrow T$ be a homomorphism, then*

(a) *If S' is a subsemigroup [submonoid] of S, then $f(S')$ is a subsemigroup [submonoid] of T.*

(b) *If T' is a subsemigroup [submonoid] of T, then $f^{-1}(T')$ is a subsemigroup [submonoid] of S.*

(c) *If $f : S \rightarrow T$ is an isomorphism, then $f^{-1} : T \rightarrow S$ is an isomorphism.*

Definition 1.40   *A nonempty subset T of a semigroup S is a **left ideal** of S if $s \in S$ and $t \in T$ implies $ts \in T$. A nonempty subset T of a semigroup S is a **right ideal** of S if $s \in S$ and $t \in T$ implies $st \in T$. A subset T of a semigroup S is an **ideal** of S if it is both a left ideal of S and a right ideal of S.*

Obviously an ideal of S is a subsemigroup of S.

Example 1.32   Let S be the semigroup of 2×2 matrices with multiplication as the operation and the integers as elements. Then matrices of the form $\begin{bmatrix} a & 0 \\ b & 0 \end{bmatrix}$ form a left ideal and matrices of the form $\begin{bmatrix} a & b \\ 0 & 0 \end{bmatrix}$ form a right ideal.

Example 1.33   The semigroup of even integers form an ideal of $(Z, \cdot)$.

Definition 1.41   *An equivalence relation $\mathcal{R}$ on a semigroup S is a **congruence** if for all $a, b, c, d \in S$, $a\mathcal{R}b$ and $c\mathcal{R}d$ imply $ac\mathcal{R}bd$.*

Definition 1.42   *Let $\mathcal{R}$ be a congruence on a semigroup S. Let $a_\mathcal{R}$ be the congruence class containing a. The set $S/\mathcal{R}$ of all the congruence classes with the multiplication $a_\mathcal{R} \cdot b_\mathcal{R} = (a \star b)_\mathcal{R}$ is called the quotient semigroup relative to the congruence $\mathcal{R}$.*

Example 1.34   Let R, the set of real numbers, be a semigroup with the operation addition [multiplication] and define $a\mathcal{R}b$ if $a - b$ is a multiple of 5. Then [0], [1], [2], [3], and [4] form a semigroup with the operation addition [multiplication].

The proof of the following theorem is left to the reader.

Theorem 1.15   *Let $\mathcal{R}$ be a congruence on a semigroup S. Then $S/\mathcal{R}$ is a semigroup with the operation defined in the previous definition and $\phi_\mathcal{R} : S \to S/\mathcal{R}$ defined by $\phi_\mathcal{R}(s) = s_\mathcal{R}$ is a homomorphism.*

Theorem 1.16   *Let $f : A \to B$ be a homomorphism and $\mathcal{R}$ be the congruence $a\mathcal{R}a'$ iff $f(a) = f(a')$, then there exists a homomorphism $g : A/_\mathcal{R} \to B$ defined by $g(a_\mathcal{R}) = f(a)$. Hence $g \circ \phi_\mathcal{R} = f$.*

Proof   We showed in Theorem 1.9 that g is a function.

$$g(a_\mathcal{R} \cdot a'_\mathcal{R}) = f(a \cdot a')$$
$$= f(a) \cdot f(a')$$
$$= g(a_\mathcal{R}) \cdot g(a'_\mathcal{R}).$$

$\square$

Theorem 1.17   *Let $f : A/_\mathcal{R} \to B$ be a function and $\mathcal{S} \subseteq \mathcal{R}$, so if $a\mathcal{S}a'$ implies $a\mathcal{R}a'$, then there exist functions $g : A/_\mathcal{S} \to B$ and $i : A/_\mathcal{S} \to A/_\mathcal{R}$ such that $f \circ i = g$.*

Proof   Let $i : A/_\mathcal{S} \to A/_\mathcal{R}$ be defined by $i(a_\mathcal{S}) = a_\mathcal{R}$ and $g : A/_\mathcal{S} \to B$ by $g(a_\mathcal{S}) = f(a_\mathcal{R})$. The function i is trivially well defined and a homomorphism.

The proof that g is a function is similar to the proof of Theorem 1.9. (See Theorem 1.10).

$$g(a_S \cdot a'_S) = g(aa'_S)$$
$$= f(aa'_R)$$
$$= f(a_R a'_R)$$
$$= f(a_R)f(a'_R)$$
$$= g(a_S)g(a'_S).$$

$\square$

We already know that the set of all functions from a set A to itself form a semigroup since for $a \in A$, and functions f, g, and h from A to itself, $((f \circ (g \circ h))(a) = ((f \circ g) \circ h)(a) = f(g(h(a))$. Also since f, g, and h are relations we have already proven that $(f \circ (g \circ h)) = (f \circ g) \circ h$.

Conversely, given a semigroup S, and $s \in S$ we can define a function ϕ_s : $S \to S$ by $\phi_s(t) = st$ for all $t \in S$. Let $T_S = \{\phi_s : S \to S$ for $s \in S\}$. For all s, t, and u in S,

$$\phi_{st}(u) = st(u)$$
$$= \phi_s(tu) = \phi_s\phi_t(u)$$
$$= (\phi_s \circ \phi_t)(u)$$

and $\phi_{st} = (\phi_s \circ \phi_t)$. Let $\tau : S \to T_S$ be defined by $\tau(s) = \phi_s$. The function τ is a homomorphism since

$$\tau(st) = \phi_{st} = \phi_s \circ \phi_t = \tau(s) \cdot \tau(t).$$

Theorem 1.18 *Every semigroup is isomorphic to a semigroup of functions from a set to itself with operation composition. If S is a monoid, then S is isomorphic to a monoid of functions from S to itself.*

Proof Given a semigroup S, and $s \in S$ define $\phi_s^1 : S^1 \to S^1$ by $\phi_s^1(t) = st$ for all $t \in S^1$ and let $T_S^1 = \{\phi_s^1 : S^1 \to S^1$ for $s \in S\}$. Let $\tau^1 : S \to T_S^1$ be defined by $\tau^1(s) = \phi_s^1$. Using the same argument as above, we see that τ^1 is a homomorphism. But if $\phi_s^1 = \phi_t^1$, since $\phi_s^1(1) = s1 = s$ and $\phi_t^1(1) = t1 = t$ then $s = t$ and τ^1 is an isomorphism. The second part of the theorem follows immediately. $\square$

Exercises

(1) Prove Theorem 1.13. Let $(S, \star)$ be a semigroup and W be a nonempty subset of S. The set of all finite products of elements of W together with the elements of W is the subsemigroup generated by W.

(2) Prove Theorem 1.14. Let $f : S \to T$ be a homomorphism, then
  (a) If S' is a subsemigroup [submonoid] of S, then $f(S')$ is a subsemigroup [submonoid] of T.
  (b) If T' is a subsemigroup [submonoid] of T, then $f^{-1}(T')$ is a subsemigroup [submonoid] of S.
  (c) If $f : S \to T$ is an isomorphism, then $f^{-1} : T \to S$ is an isomorphism.
(3) Prove Theorem 1.15. Let $\mathcal{R}$ be a congruence on a semigroup S. Then $S/\mathcal{R}$ is a semigroup with the operation defined by $a_{\mathcal{R}} \cdot b_{\mathcal{R}} = (a \star b)_{\mathcal{R}}$ and $\phi_{\mathcal{R}} : S \to S/\mathcal{R}$ is a homomorphism.
(4) Prove that a finite semigroup S contains a subgroup.
(5) Give an example of a group which contains a subsemigroup that is not a monoid.
(6) Prove that the identity of a monoid is unique.
(7) Prove that if a semigroup contains a 0, then it is unique.
(8) Prove that if G is a finite group and H is a subgroup of G, then $|H| = |gH|$ for every $g \in G$.
(9) Prove that if G is a finite group and H is a subgroup of G, then $H = gH$ if and only if $g \in H$.
(10) Prove that if G is a finite group and H is a subgroup of G, then $|H|$ divides $|G|$.
(11) An idempotent of a semigroup S is an element a such that $a \cdot a = a$. Prove that if $f : S \to T$ is a homomorphism, then if a is an idempotent, $f(a)$ is an idempotent.
(12) An element a of a semigroup S is a **left identity** if $as = s$ for all $s \in S$. An element a of a semigroup S is a **right identity** if $sa = s$ for all $s \in S$. Give an example of a semigroup having more that one left identity.
(13) Let $f : S \to T$ be a homomorphism. Prove that if T contains 0, then $f^{-1}(0)$ is an ideal.
(14) Using the properties in Theorem 1.1, prove that if $S = \mathcal{P}(C)$ for some nonempty set C, then $(S, \triangle)$ is a monoid, where $\triangle$ denotes the symmetric difference. What is the identity? Is $(S, \triangle)$ a group?
(15) Let $S = \mathcal{P}(C)$ for some nonempty set C, is $(S, \cup)$ a monoid? Is $(S, \cap)$ a monoid? Are they groups?
(16) Define the multiplication of permutations of a set to be composition as shown in the previous section. Prove that the set of permutations of a set with this multiplication form a group.

2

Languages and codes

2.1 Regular languages

Definition 2.1 *An **alphabet**, denoted by Σ, is a set of symbols. A **string** or **word** is a sequence $a_1a_2a_3a_4\ldots a_n$ where $a_i \in \Sigma$.*

Thus if $\Sigma = \{a, b\}$, then aab, a, $baba$, $bbbbb$, and $baaaaa$ would all be strings of symbols of Σ. In addition we include an empty string denoted by λ which has no symbols in it.

Definition 2.2 *Let Σ^* denote the set of all strings of Σ including the empty string. Define the binary operation $\circ$ called **concatenation** on Σ^* as follows: If $a_1a_2a_3a_4\ldots a_n$ and $b_1b_2b_3b_4\ldots b_m \in \Sigma^*$ then*

$$a_1a_2a_3a_4\ldots a_n \circ b_1b_2b_3b_4\ldots b_m = a_1a_2a_3a_4\ldots a_nb_1b_2b_3b_4\ldots b_m.$$

If S and T are subsets of Σ^* then $S \circ T = \{s \circ t : s \in S, t \in T\}$. The set $S \circ T$ is often denoted as ST.

Thus if $\Sigma = \{a, b\}$, then $aabba \circ babaa = aabbababaa$. In particular, if ω is a string in Σ^* then $\lambda \circ \omega = \omega \circ \lambda = \omega$, so that a string followed or preceded by the empty string simply gives the original string. Notice that in general it is not true that $w \circ v = v \circ w$.

The following is a specific case of the submonoid generated by a subset of a monoid described in Chapter 1.

Definition 2.3 *Let B be a subset of Σ^* then B^* is the set of all strings or words formed by concatenating words from B together with the empty string, i.e. $B^* = \{w_1w_2\ldots w_n : w_i \in B\} \cup \{\lambda\}$. If $\emptyset$ denotes the empty set then $\emptyset^* = \{\lambda\}$.*

The symbol * is called the **Kleene star** and is named after the mathematician and logician Stephen Cole Kleene.

Note that Σ^* is consistent with this definition.

23

Example 2.1 $\{a\}^* = \{\lambda, a, aa, aaa, \ldots\}$.

Example 2.2 $\{a\}\{ab\}^*\{c\} = \{ac, aabc, aababc, \ldots\}$.

Let A^+ be the set consisting of all finite products of elements of a nonempty set A together with the operation of concatenation. The set $A^+ = \langle A \rangle$, and hence is a semigroup as shown in Theorem 1.13. From Theorem 1.13 and the definition of A^* we know that if A is a nonempty subset of Σ then the set $(A^*, \circ)$ is a monoid where λ is the identity. It is the **submonoid generated** by A. If A does not contain the empty word, then $(A^+, \circ)$ differs from $(A^+, \circ)$ since it contains the empty word. Thus if $A = \{a\}$, then $A^+ = \{a, a^2, a^3, \ldots\}$ and $A^* = \{\lambda, a, a^2, a^3, \ldots\}$. Note that $A^+ = aA^*$.

Definition 2.4   *Let Σ^* denote the set of all strings of Σ including the empty string. A subset L of Σ^* is called a **language**.*

If Σ is the set of letters in the English alphabet, then L could be the set of words in the English language. If Σ is the set of letters in the Greek alphabet, then L could be the set of words in the Greek language. If Σ is the set of symbols used in a computer language, then L could be the set of words in that language. Since every subset of Σ^* is a language, many will be difficult or impossible to describe. In particular a language is not necessarily closed under the operation of concatenation.

If Σ is the set $\{a, b, c\}$ then the following are languages:

$L_1 = \{a, aab, aaabb, aaaabbb \ldots\}$,
$L_2 = \{w : w \in \Sigma^* \text{ and contains exactly one } a \text{ and one } b\}$,
$L_3 = \{w : w \in \Sigma^* \text{ and contains exactly two } bs\}$,
$L_4 = \{w : w \in \Sigma^* \text{ and contains at least two } bs\}$,
$L_5 = \{w : w \in \Sigma^* \text{ and contains the same number of } as, bs, \text{ and } cs\}$,
$L_6 = \{w : w = a^n b^n \text{ for } n \geq 1\}$,
$L_7 = \{w : w = a^n b^n c^n \text{ for } n \geq 1\}$,
$L_8 = \{w : w \in \Sigma^* \text{ and contains no } cs\}$.

Definition 2.5   *Let Σ be an alphabet. The class of **regular expressions R** over Σ is defined by the following rules using Σ and the symbols $\emptyset$, λ, *, $\vee$, $(, and)$. The symbol λ is used to denote the symbol $\emptyset^*$.*

  (i) *The symbol $\emptyset$ is a regular expression and for every $a \in \Sigma$, the symbol **a** is a regular expression.*
 (ii) *If w_1 and w_2 are regular expressions, then $w_1 w_2$, $w_1 \vee w_2$, w_1^*, and (w_1) are regular expressions.*
(iii) *There are no regular expressions which are not generated by (i) and (ii).*

Each expression corresponds to a set with the following correspondence £ defined by

$$£(\emptyset) \quad = \emptyset,$$
$$£(\mathbf{a}) \quad = \{a\} \text{ for all } a \in \Sigma,$$
$$£(\lambda) \quad = \{\lambda\},$$
$$£(w_1 \vee w_2) = £(w_1) \cup £(w_2) \text{ for expressions } w_1, w_2,$$
$$£(w_1 w_2) \quad = £(w_1) \circ £(w_2),$$
$$£(w_1^*) \quad = £(w_1)^*,$$

so that

$£(\mathbf{aa}^*)$	$= \{a\} \circ \{a\}^*$	$= \{a, aa, aaa, aaaa, aaaa, \ldots\},$
$£(\mathbf{a}(\mathbf{b} \vee \mathbf{c})\mathbf{d})$	$= \{a\} \circ \{\{b\} \cup \{c\}\} \circ \{d\}$	$= \{abd, acd\},$
$£((\mathbf{a} \vee \mathbf{b})^*)$	$= \{a \cup b\}^*$	$= \{\lambda, a, b, ab, ba, abb, aba, \ldots\}$
		$= \text{all strings consisting of}$
		$\text{0 or more } as \text{ and } bs,$
$£(\mathbf{ab}^*\mathbf{c})$	$= \{a\} \circ \{b\}^* \circ \{c\}$	$= \{ac, abc, abbc, abbbc, \ldots\},$
$£(\mathbf{a}^* \vee \mathbf{b}^* \vee \mathbf{c}^*)$	$= \{a\}^* \cup \{b\}^* \cup \{c\}^*$	$= \{\lambda, a, b, c, \ldots, a^k, b^k, c^k \ldots\},$
$£(\lambda)$	$= £(\emptyset^*)$	$= \{\lambda\},$
$£((\mathbf{a} \vee \mathbf{b})\mathbf{c}))$	$= (\{a\} \cup \{b\}) \circ \{c\}$	$= \{ac, bc\}.$

The image of a regular expression is a regular language. Regular languages may be defined as follows:

Definition 2.6 *The class R of a **regular languages** over Σ has the following properties:*

  (i) *The empty set, $\emptyset \in R$, and if $a \in \Sigma$, then $\{a\} \in R$.*
 (ii) *If s_1 and $s_2 \in R$, then $s_1 \cup s_2$, $s_1 \circ s_2$, $s_1^* \in R$.*
(iii) *Only sets formed using (i) and (ii) belong to R.*

Although it will not be shown until later, the intersection of regular sets is a regular set and the complement of a regular set is a regular set.

The previous definitions of regular languages and regular expressions are examples of **recursive definitions**. In a recursive definition there are three steps. (1) Certain objects are defined to be in the set. (2) Rules are listed for producing new objects in the set from other objects already in the set. (3) There are no other elements in the set. Mathematical induction is a special case of a recursive definition. We shall see that the set $\{ab, a^2b^2, \ldots, a^nb^n, \ldots\}$ is not a regular set. However, we cannot assume this is the case because we cannot immediately describe the set using the definition. In general it is not always easy to show that a set is not regular. Later, we shall show how to determine that many sets are not regular.

Example 2.3 Examples of regular expressions include *(a ∨ b)**, *(a* ∨ b *)*, *a* (c ∨ d) a (b ∨ a ∨ c)**, and λ. Examples of regular sets include $\{a, b, c\}$, $\{a\}^*$, $\{ab\}^*$, $\{c\}\{b\}^*$, $\{a\} \vee \{b\} \vee \{cd\}$, and $(\{a\} \vee \{b\})^* \vee \{c^*d\} \vee \{\lambda\}$.

As mentioned previously, not all classes of languages are so easily defined. In the following chapters we shall define machines that generate languages and machines that accept languages. A machine accepts a language if it can determine whether a string is in the language. Many languages are defined by the fact that they can be generated or accepted by a particular type of machine.

If $T^* = S$, T is not usually uniquely defined. If $T = \{a, b, c, d\}$, and $\overline{T} = \{a, b, c, d, ab, cd, bc\}$, then $T^* = \overline{T}^*$ but, while every string in T^* can be expressed uniquely as the concatenation of elements of T, this is not true of elements of $\overline{T}$ since the expression **abcd** can be expressed as **(a)(b)(c)(d)**, and also as **(a)(b)(cd)**, **(a)(bc)(d)**, **(ab)(cd)**, etc.

Definition 2.7 *A **code** is a subset of Σ^*. If C is a subset of Σ^* and every string in S can be expressed as the concatenation of elements of C, then we say that C is a code for S. A code C is **uniquely decipherable** if every string in S can be uniquely expressed as the concatenation of elements of C.*

Therefore $\{ba, ab, ca\}$, $\{ade, ddbee, dfc, dgd\}$, and $\{ae, b, c, de\}$ are uniquely decipherable codes while $\{a, ab, bc, c\}$, $\{ab, abc, cde, de\}$, and $\{a, bc, ab, c\}$ are not uniquely decipherable codes.

Note that in many texts, a subset of Σ^* is defined to be a code only if it is uniquely decipherable.

Definition 2.8 *Let Σ be an alphabet. A nonempty code $C \subseteq \Sigma^*$ is called a **prefix code** if for all words $u, v \in C$, if $u = vw$ for $w \in \Sigma^*$, then $u = v$ and $w = \lambda$. This means that no word in a code can be the beginning string of another word in the code. A nonempty code $C \subseteq \Sigma^*$ is called a **suffix code** if for all words $u, v \in C$, if $u = wv$ for $w \in \Sigma^*$, then $u = v$ and $w = \lambda$. This means that no word in a code can be the final string of another word in the code. A nonempty code $C \subseteq \Sigma^*$ is called a **biprefix code** if it is both a prefix and a suffix code. A nonempty code $C \subseteq \Sigma^*$ is called an **infix code** if no word in the code can be a substring of another word in the code so that if u and wuv are words in the code for $w, v \in \Sigma^*$, then $w = v = \lambda$. A code is called a **block code** if each string in the code has equal length.*

The set $\{a, ab, abc\}$ is a uniquely decipherable code but it is not a prefix code since a is the initial string of both ab and abc and ab is an initial string of abc. It is however a suffix code. The set $\{a, ba, ca\}$ is a prefix code, but it is not a suffix code since a is the final string of both ba and ca. The set $\{ad, ab, ac\}$ is a biprefix code. Any code whose regular expression begins with a^* is not a suffix code and any code whose regular expression ends with a^* is not a prefix code. However, a^*b is a prefix code and ab^* is a suffix code.

The proofs of the following theorems are left to the reader:

Theorem 2.1 *If a code is a suffix, prefix, infix, biprefix, or block code, then it is uniquely decipherable.*

Theorem 2.2 *A block code is a suffix, prefix, infix, and biprefix code.*

Theorem 2.3 *An infix code is a biprefix code.*

Exercises

(1) Given $w = 10110$, find five words v_1, v_2, v_3, v_4, v_5 such that $v_i w = w v_i$ for $1 \le i \le 5$.

(2) Find regular sets corresponding to the following expressions. If the set is infinite, list ten elements in the set:
  (a) $\mathbf{a}(\mathbf{b} \vee \mathbf{c} \vee \mathbf{d})\mathbf{a}$
  (b) $\mathbf{a}^*\mathbf{b}^*\mathbf{c}$
  (c) $(\mathbf{a} \vee \mathbf{b})(\mathbf{c} \vee \mathbf{d})$
  (d) $(\mathbf{ab}^*\lambda) \vee (\mathbf{cd})^*$
  (e) $\mathbf{a}(\mathbf{bc})^*\mathbf{d}$.

(3) Find regular sets corresponding to the following expressions. If the set is infinite, list ten elements in the set:
  (a) $\mathbf{bc}(\mathbf{bc})^*$
  (b) $(\mathbf{a} \vee \mathbf{b}^* \vee \lambda)(\mathbf{c} \vee \mathbf{d}^*)$
  (c) $(\mathbf{a} \vee \mathbf{bc} \vee \mathbf{d})^*$
  (d) $(\mathbf{a} \vee \mathbf{b})(\mathbf{c} \vee \mathbf{d})\mathbf{b}$
  (e) $\mathbf{a}^*(\mathbf{b} \vee \mathbf{c} \vee \mathbf{d})^*$.

(4) Find regular expressions that correspond to the following regular sets:
  (a) $\{ab, ac, ad\}$
  (b) $\{ab, ac, bb, bc\}$
  (c) $\{a, ab, abb, abbb, abbbb, \ldots\}$
  (d) $\{ab, abab, ababab, abababab, ababababab, \ldots\}$
  (e) $\{ab, abb, aab, aabb\}$.

(5) Find regular expressions that correspond to the following regular sets:
  (a) $\{ab, acb, adb\}$
  (b) $\{ab, abb, abbb, abbbb, \ldots\}$
  (c) $\{ad, ae, af, bd, be, bf, cd, ce, cf\}$
  (d) $\{abcd, abcbcd, abcbcbcd, abcbcbcbcd, \ldots\}$
  (e) $\{abcd, abef, cdcd, cdef\}$.

(6) Let $\Sigma = \{a, b, c\}$.

    (a) Give a regular expression for the set of all elements of Σ^* containing exactly two bs

    (b) Give a regular expression for the set of all elements of Σ^* containing exactly two bs and two cs

    (c) Give a regular expression for the set of all elements of Σ^* containing two or more bs

    (d) Give a regular expression for the set of all elements of Σ^* beginning and ending with a and containing at least one b and one c.

    (e) Give a regular expression for the set of all elements of Σ^* consisting of one or more as, followed by one or more bs and then one or more cs.

(7) Let $\Sigma = \{a, b\}$.

    (a) Give a regular expression for the set of all elements of Σ^* containing exactly two bs or exactly two as.

    (b) Give a regular expression for the set of all elements of Σ^* containing an even number of bs.

    (c) Give a regular expression for the set of all elements of Σ^* beginning and ending with a and containing at least one b.

    (d) Give a regular expression for the set of all elements of Σ^* such that the number of as in each string is divisible by 3 or the number of bs is divisible by 5.

    (e) Give a regular expression for the set of all elements of Σ^* such that the length of each string is divisible by 3.

(8) Which of the following are uniquely decipherable codes?

    (a) $\{ab, ba, a, b\}$

    (b) $\{ab, acb, accb, acccb, \ldots\}$

    (c) $\{a, b, c, bd\}$

    (d) $\{ab, ba, a\}$

    (e) $\{a, ab, ac, ad\}$.

(9) Which of the following expressions describe uniquely decipherable codes?

    (a) **ab***

    (b) **ab***$\vee$ **baaa**

    (c) **ab***$\mathbf{c} \vee$ **baaac**

    (d) $(\mathbf{a} \vee \mathbf{b})(\mathbf{b} \vee \mathbf{a})$

    (e) $(\mathbf{a} \vee \mathbf{b} \vee \lambda)(\mathbf{b} \vee \mathbf{a} \vee \lambda)$.

(10) Which of the following are uniquely decipherable codes? Which are suffix codes?

    (a) $\{ab, ba\}$

    (b) $\{ab, abc, bc\}$

    (c) $\{a, b, c, bd\}$

    (d) $\{aba, ba, c\}$

    (e) $\{ab, acb, accb, acccb\}$.

(11) Which of the following expressions describe prefix codes? Which describe suffix codes?

    (a) **ab***

    (b) **ab***c**

    (c) **a***bc***

    (d) **(a ∨ b)(b ∨ a)**

    (e) **a***b**.

(12) Show that the intersection of the monoids $\{ac, bc, d\}^*$ and $\{a, cb, cd\}^*$ is the monoid generated by the code described by the expression **ac(bc)*d**.

(13) Prove Theorem 2.1. If a code is a suffix, prefix, infix, biprefix, or block code, then it is uniquely decipherable.

(14) Prove Theorem 2.2. A block code is a suffix, prefix, infix, and biprefix code.

(15) Prove Theorem 2.3. An infix code is a biprefix code.

2.2 Retracts (Optional)

In this section we discuss an additional source of examples of regular languages: the fixed languages of endomorphisms of free monoids A^*. Each such language is necessarily a submonoid of A^* and is the image of a special type of endomorphism called a retraction. Such images are called retracts and they are characterized among submonoids as those submonoids that are generated by a special class of codes called key codes.

Definition 2.9 *Let X be a set and let $f : X \to X$ be a function having the property that $f(f(x)) = f(x)$ for all x in X. A function with this property is called a **retraction** of X and its image is called a **retract** of X.*

    Notice that the restriction of a retraction $f : X \to X$ to the image of f, $\{f(x) : x \in X\}$, is the identity mapping of the image of f onto itself.

Example 2.4 For the real numbers R, the absolute value function $f : R \to R$, defined by $f(r) = |r|$, is a retraction and $\{x \in R : x \geq 0\}$ is its associated retract. The floor and ceiling functions, when regarded as functions from R into R provide two additional examples of retractions which determine the same retract $\{r \in R : r$ is an integer$\}$.

Notice that, when X is merely a set having no specified structure, every nonempty subset S of X is a retract of S since S is the image of the retraction $r : X \to X$ defined by $r(x) = x$ if x is in S and otherwise $r(x) = s$ where s is in S. When structures are specified on a set, its retracts may become quite interesting. In this section we study the retracts of free monoids A^*, where A is a finite set. Several of our results hold also when A is infinite but we leave these extensions to the interested reader. We will assume in this section and the next that alphabets are always finite.

Definition 2.10   *The fixed language of a homomorphism* $h : A^* \to A^*$ *is the set* $L = \{w \in A^* : h(w) = w\}$.

Note that the fixed language of each homomorphism is a submonoid of A^*.

Example 2.5   Let $A = \{a, b, c, d\}$ and let f be the homomorphism $f : A^* \to A^*$ defined by $f(a) = dad$, $f(b) = bc$, $f(c) = d$, $f(d) = \lambda$. The fixed language of f is the submonoid generated by the set $\{dad, bcd\}$. Notice that $f(f(b)) = f(bc) = f(b)f(c) = bcd$ which is not equal to $f(b) = bc$. Consequently f is not a retraction. However, notice also that the homomorphism $r : A^* \to A^*$, defined by $r(a) = dad, r(b) = bcd, r(c) = r(d) = \lambda$ is a retraction and has the same image as f. Thus the image of f is a retract even though f itself is not a retraction. Finally, note that $\{dad, bcd\}$ is a uniquely decipherable code.

The behavior of a, b, and c in Example 2.5 will provide an illustration of the classification of alphabetical symbols that will be necessary for understanding retracts.

Definition 2.11   *Let A be a set and let f be a homomorphism* $f : A^* \to A^*$. *A symbol a in A is said to be **mortal**, with respect to f, if there is a positive integer n for which $f^n(a) = \lambda$; otherwise a is said to be **vital**.*

For each homomorphism f, the mortal/vital dichotomy of the symbols of A may be determined as follows. For each nonnegative integer j let A_j be defined inductively by: A_0 is empty; $A_1 = \{a \in A : h(a) = \lambda\}$; and for $j \geq 2$, $A_j = \{a \in A : h(a) \in A_{j-1}^*\}$. Since A is finite there will be a least nonnegative integer m for which $A_m = A_{m+1}$. The set A_m is the set of all mortal symbols and its complement in A is the set of vital symbols.

Notice that in Example 2.5 the symbols d and c are mortal and the symbols a and b are vital. Note also that the fixed language is the submonoid generated by a set of words in each of which there is exactly one occurrence of a vital symbol. Further, each of these vital symbols occurs in only one of the generators. In this section we show that the simple observations concerning fixed languages, retractions, and codes, made for Example 2.5, are completely typical of fixed languages of homomorphisms.

Definition 2.12 *Any symbol k in A that occurs exactly once in a word w in A^* is called a* **key** *of w. A word w for which there is at least one key symbol is called a* **key word**.

Note that for the word *dad* the symbol *a* is the unique key. For the word *bcd* each of the three symbols *b*, *c*, and *d* is a key. Consequently both *dad* and *bcd* are key words. Finally, the word *abcbac* is not a key word since it has no key.

Definition 2.13 *A set X of words is called a* **key code** *if each word in X is a key word and a key for each word in X can be chosen that does not occur in any other word in X.*

Note that the set of generators given in Example 2.5, namely $X = \{dad, bcd\}$, is a key code. The word *dad* allows only the unique symbol *a* to be chosen as its key. The word *bcd* allows each of *b*, *c*, and *d* to be chosen as a key. To confirm that *X* is a key code we cannot use *d* as a key for the word, but if either *b* or *c* is chosen as the key for *bcd*, then it is confirmed that *X* is a key code. Each key code *X* is uniquely decipherable since, given any string that is the concatenation of words chosen from *X*, simply noting the key symbols that occur in the string provides the unique segmentation of *X* into code words. The key codes constitute a very restricted subclass of the uniquely decipherable codes. Such simple codes as $\{aa, bb, cc, dd\}$ are uniquely decipherable, but contain no key word at all.

Example 2.6 Let $A = \{a, b, c, d\}$. The following are key codes: $\{a, b, c, d\}$, $\{a, bcc, dcc\}$, $\{abcbc, bbd\}$, $\{ababcd\}$, and the empty set. Note the crucial fact that $\{a, b, c, d\}$ is the *only key code* in A^* that consists of exactly *four* words, since each four-word key code must use each of the four symbols in *A* as a key. The following are not key codes: $\{abba\}$, $\{abcd, c\}$, $\{abc, bcd, cda\}$, $\{a, b, c, dd\}$. Note that a subset of A^* that contains *five* or more words cannot be a key code, since there are only four possible keys.

The following technical result is the basis for the theorem that establishes the firm relationships that hold among the concepts: fixed language, retract, and key code.

The following proposition was discovered by Tom Head [16].

Proposition 2.1 *Let A be an alphabet and $h : A^* \to A^*$ be a homomorphism. Let $X = \{a \in A : h(a) = uav$ where only mortal symbols occur in u and $v\}$. For each a in X, let N_a be the least nonnegative integer for which $h^{N_a}(uv) = \lambda$. Let $H = \{h^{N_a}(a) : a \in X\}$. The fixed language L of h is the submonoid of A^**

generated by H. The correspondence $a \leftrightarrow h^{N_a}(a)$ is a one-to-one correspondence between X and H.

Proof (1) $H^* \subseteq L$: Since h is a homomorphism it is sufficient to verify that each element $h^{N(a)}(a)$ of H is in L, which is confirmed by the calculation:

$$
\begin{aligned}
h(h^{N(a)}(a)) &= h^{N(a)}(h(a)) \\
&= h^{N(a)}(uav) \\
&= h^{N(a)}(u)h^{N(a)}(a)h^{N(a)}(v) \\
&= \lambda h^{N(a)}(a)\lambda \\
&= h^{N(a)}(a).
\end{aligned}
$$

(2) $L \subseteq H^*$: Let w be in L. Let $a_1, a_2, a_3, \ldots, a_n$ be the subsequence consisting of the occurrences of the vital elements of w. We use now the principle: a vital symbol can come only from a vital symbol and only a mortal symbol can eventually be erased. Since $h(w) = w$, each a_i must occur exactly once in $h(a_i)$. It follows that, for each a_i, we must have $h(a_i) = u_i a_i v_i$ where each u_i, v_i must consist entirely of mortal symbols. Thus for each i there is a least nonnegative integer $N(i)$ for which $h^{N(i)}(u_i v_i) = \lambda$. Let N be the largest of the $N(i)$. Note that, for each a_i, $h^N(a_i) = h^{N(i)}(a_i)$ is in H. Then $w = h(w) = h^N(w) = h^N(a_1)h^N(a_2)h^N(a_3)\ldots h^N(a_n)$ is in H^*.

From (1) and (2) we have $L = H^*$. Note that $H = \{h^{N(i)}(a) : a \in X\}$ is a key code since the set X is a set of keys for H. $\qquad\square$

Theorem 2.4 *Let A be a finite alphabet and L be a language contained in A^*. The following three conditions on L are equivalent:*

(1) L is the fixed language of a homomorphism of A^* into A^*;
(2) L is a submonoid of A^* that is generated by a key code; and
(3) L is a retract of A^*.

Proof (1 $\Rightarrow$ 2): Let $h : A^* \to A^*$ be a homomorphism. Proposition 2.1 provides us with the key code H for which $L = H^*$.

(2 $\Rightarrow$ 3): Let L be a submonoid of A^* that is generated by a keycode X and let K be a set of keys for X. For each k in K there are strings x_k and y_k for which $x_k k y_k$ is the key word in X having k as its key. Define a homomorphism $r : A^* \to A^*$ by $r(k) = x_k k y_k$ for each k in K and $r(a) = \lambda$ for each a not in K. Note that r is a retraction of A^* having $X^* = L$ as its image, hence L is a retract.

(3 $\Rightarrow$ 1): Let L be a retract of A^*. Then there is a retraction $r : A^* \to A^*$ that has L as its image. Then L is the fixed language of the the homomorphism r. $\qquad\square$

Theorem 2.4 has several valuable corollaries the proofs of which will be relegated to the exercises.

Corollary 2.1 *A retract of a free monoid is free.*

Note that this property of retracts does not hold for arbitrary submonoids of a free monoid since, for any nonempty alphabet A, the submonoid consisting of all words of length ≥ 2 is *not* free.

Corollary 2.2 *If A is an alphabet having exactly n symbols, then no inclusion chain of distinct retracts of A^* has more than $n + 1$ retracts even when the retract $\{\lambda\}$ is included.*

Corollary 2.3 *If X is a key code and x^n lies in X^*, then so does x.*

Corollary 2.4 *If X is a key code and both uv and vu lie in X^*, then so do u and v.*

Let $A = \{a_1, a_2, a_3, \ldots, a_n\}$. A simple example of a longest possible inclusion chain of retracts in A^* is

$$\{a_1, a_2, a_3, \ldots, a_n\}^*, \{a_2, a_3, \ldots, a_n\}^*, \{a_3, \ldots, a_n\}^*, \ldots, \{a_n\}^*, \{\lambda\}.$$

Each of these retracts, except the first, is maximal among the retracts contained in its predecessor. In each case the number of generators of the subretract is one less than the number of generators of its predecessor. However, maximal proper subretracts of a retract can have *many* fewer generators:

Proposition 2.2 *Let n be a positive integer and $A = \{a_1, a_2, a_3, \ldots, a_n\}$ an alphabet of n symbols. Let m be any positive integer less than n. Then A^* contains a maximal proper retract generated by exactly m words.*

Proof The set of m words

$$K = \Big\{a_1, a_2, a_3, \ldots, a_{m-1}, a_m^2 a_{m+1}^2 a_{m+2}^2 a_{m+3}^2 \cdots$$
$$a_{n-1}^2 a_n a_m a_{m+1} a_{m+2} a_{m+3} \cdots a_{n-1} a_n\Big\}$$

is a key code for which K^* is a maximal proper retract of A^*.

The verification of the maximality is left as an exercise. □

The retracts of a free monoid and the the partially ordered set they form under set inclusion have been studied previously in [16],[10], and [9].

Exercises

(1) Which of the following are sets of key codes?

    (a) $\{a, ab, ac, d\}$

    (b) $\{ab, ac, ad, ae\}$

    (c) $\{aabaa, aacaa, ddeda, dadaf\}$

    (d) $\{abba, acca, adda, aeea\}$.

(2) Define the retraction maps with the following retracts on A^* where $A = \{a, b, c, d, x, y, z\}$

    (a) $\{aabaa, acaax, daaxy\}$

    (b) $\{ax, bx, cx, dx\}$

    (c) $\{abcd\}$.

(3) Prove that the restriction of a retraction $f : X \to X$ to the image of f, is the identity mapping of the image of f onto itself.

(4) Prove Corollary 2.1 that a retract of a free monoid is free.

(5) Prove Corollary 2.2 – if A is an alphabet having exactly n symbols, then no inclusion chain of distinct retracts of A^* has more than $n + 1$ retracts even when the retract $\{\lambda\}$ is included.

(6) Prove Corollary 2.3 – if X is a key code and x^n lies in X^*, then so does x.

(7) Prove Corollary 2.4 – if X is a key code and both uv and vu lie in X^*, then so do u and v.

2.3 Semiretracts and lattices (Optional)

The intersection of two retracts of the free monoid on a finite set A need not be a retract if A contains four or more symbols. Possibly the simplest example is the following one adapted from [7]: Let $A = \{a, b, c, d\}$. The sets $\{ab, ac, d\}$ and $\{ba, c, da\}$ are key codes and consequently the submonoids R and R' that they generated are retracts of A^*. However, their intersection

$$R \cap R' = (d(ab)^*ac)^*$$

is not only not a retract; it is not even finitely generated. The set $d(ab)^*ac$ is a uniquely decipherable code since simply noting the locations of the occurrences of the symbol d in any string that is a concatenation of these words provides the unique segmentation of the string into generators. Thus $R \cap R'$ is a free submonoid, although not a retract of A^*. In fact, the intersection of any family, whether finite or infinite, of free submonoids of a free monoid is free [37]. Consequently the family of free submonoids of a free monoid is always a complete lattice. By broadening our attention slightly we obtain a similarly attractive stability result for what we call semiretracts of free monoids:

Definition 2.14 *By a **semiretract** of A^*, we mean an intersection of a finite number of retracts of A^*.*

Each retract of A^* is also a semiretract. The clearest example of a semiretract that is not a retract is the example given previously: $R \cap R' = (d(ab)^*ac)^*$. Some pairs of retracts have as their intersection a retract:

$$\{abc, d\}^* \cap \{a, bcd\}^* = \{abcd\}^*.$$

As stated above, but not to be demonstrated here, if fewer than four alphabet symbols appear in the keycodes that generate a class of retracts, then the intersection of this collection must also be a retract.

Since every retract of A^* is a regular language, every semiretract is also a regular language. Thus this section continues to yield examples of regular languages.

The definition of a semiretract provides one closure property, "the intersection of a finite number of semiretracts is semiretract." A stronger result is true, but not obvious, "the intersection of any *finite or infinite collection* of semiretracts of a free monoid A^* on a finite alphabet A is again a semiretract." This is an immediate consequence of a co-compactness property of the family of semiretracts of Λ^* which is included in the appendices. *Every collection of retracts of A^* has a finite sub-collection that has the same intersection as the original collection.* The intersection of the finite sub-collection is a semiretract of A^* by the definition of semiretract.

The elementary set theoretic union of two semiretracts need not be a semiretract nor even a submonoid: a^* and b^* are retracts, but $a^* \cup b^*$ is not a submonoid. However, given any collection C of semiretracts, whether finite or infinite, let M be the intersection of the class of all semiretracts of A^*, each of which contains every semiretract in C. There is at least one semiretract that contains them all, namely A^* itself. The resulting intersection M is a semiretract of A^* as explained in the previous paragraph. For each such C we denote M by $\vee C$. We summarize the discussions of this section as:

Theorem 2.5 *Let A be a finite alphabet. The set of all semiretracts of A^* is a complete lattice with binary operations $\cap$ and $\vee$ having A^* as maximal element and $\{\lambda\}$ as minimal element.*

The semiretracts of a free monoid and the lattice they form have been studied previously in [**1**], [**2**], and [**3**].

Exercises

(1) Find the code of the semiretract which is the intersection of retracts with key codes $\{ab, cb, cd\}$ and $\{a, bc, d\}$.

(2) Find the code of the semiretract which is the intersection of retracts with key codes $\{ab, st, sd, ef, eg\}$ and $\{a, bs, ts, de, fe, g\}$.

(3) Find the code of the semiretract which is the intersection of retracts with key codes $\{ba, st, sd, ef, eg\}$ and $\{as, bs, ts, de, fe, g\}$.

(4) Find the key codes of two retracts whose intersection has the basis $ab(def)^*dgh$.

(5) Find the key codes of two retracts whose intersection has the basis $ab(de)^*dfg(hk)^*hm$.

(6) Prove that a key code of a retract is a prefix code.

(7) Prove that a key code of a retract is an infix code.

(8) Find a semiretract that is the intersection of three retracts, but not two retracts.

3

Automata

3.1 Deterministic and nondeterministic automata

An **automaton** is a device which recognizes or accepts certain elements of Σ^*, where Σ is a finite alphabet. Since the elements accepted by the automaton are a subset of Σ^*, they form a language. Therefore each automaton will recognize or accept a language contained in Σ^*. The language of Σ^* consisting of the words accepted by an automaton M is the **language over Σ^* accepted by** M and denoted $M(L)$. We will be interested in the types of language an automaton accepts.

Definition 3.1 *A **deterministic automaton**, denoted by $(\Sigma, Q, s_0, \Upsilon, F)$, consists of a finite alphabet Σ, a finite set Q of states, and a function $\Upsilon : Q \times \Sigma \to Q$, called the **transition function** and a set F of acceptance states. The set Q contains an element s_0 and a subset F, the set of **acceptance states**.*

The input of Υ is a letter of Σ and a state belonging to Q. The output is a state of Q (possibly the same one). If the automaton is in state s and "reads" the letter a, then (s, a) is the input for Υ and $\Upsilon(s, a)$ is the next state. Given a string in Σ^* the automaton "reads" the string or word as follows. Beginning at the initial state s_0, and beginning with the first letter in the string (if the string is nonempty), it reads the first letter of the string. If the first letter is the letter a of Σ, then it "moves" to state $s = \Upsilon(s_0, a)$. The automaton next "reads" the second letter of the string, say b, and then moves to state $s' = \Upsilon(s, b)$. Therefore, as the automaton continues to "read" a string of letters from the alphabet it "moves" from one state to another. Eventually the automaton "reads" every letter in the string and then stops. If the state the automaton is in after reading the last letter belongs to the set of acceptance states, then the automaton **accepts** the string. Let M be the automaton with alphabet $\Sigma = \{a, b\}$, set of states $Q = \{s_0, s_1, s_2\}$,

and Υ defined by the table

Υ	s_0	s_1	s_2
a	s_1	s_2	s_2
b	s_0	s_0	s_1

Suppose M "reads" the string aba. Since the automaton begins in state s_0, and the letter read is a, and $\Upsilon(s_0, a) = s_1$, the automaton is now in state s_1. The next letter read is b and $\Upsilon(s_1, b) = s_0$. Finally the last letter a is read and, since $\Upsilon(s_0, a) = s_1$, the automaton remains in state s_1. We may also state Υ as a set of rules as follows:

> If in state s_0 and a is read go to state s_1.
> If in state s_1 and a is read go to state s_2.
> If in state s_2 and a is read go to state s_2.
> If in state s_0 and b is read go to state s_0.
> If in state s_1 and b is read go to state s_0.
> If in state s_2 and b is read go to state s_1.

Let s_0 and s_2 be the acceptance states.

This deterministic automaton is best shown pictorially by a **state diagram** which is a directed graph where the states are represented by the vertices and each edge from s to s' is labeled with a letter, say a, of the alphabet Σ if $\Upsilon(s, a) = s'$. A directed arrow from s to s' labeled with the letter a will be called an a-**arrow** from s to s'. If s is a starting state, then its vertex is denoted by the diagram

If s is an acceptance state, its vertex is denoted by the diagram

Therefore the deterministic automaton above may be represented pictorially as seen in Fig. 3.1. More specifically, an automaton "reads" a word or string $a_0 a_1 a_2 \ldots a_n$ of Σ^* by first reading a_0, then reading a_1 and continuing until it has read a_n. If an automaton is in state s_1 and reads the word w and is then in s_2, then w is a path from s_1 to s_2. A deterministic automaton **accepts** or **recognizes** $a_0 a_1 a_2 \ldots a_n$ if after beginning with a_0 in state s_0 and continuing until reading a_n, the automaton stops in an acceptance state. Thus the automaton

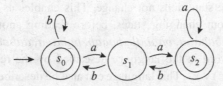

Figure 3.1

above would not accept *aba* since s_1 is not an acceptance state. It would however accept *bbaaa* and *bab*, since s_0 and s_2 are acceptance states.

The automaton with the state diagram

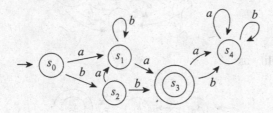

has initial state s_0 and acceptance state s_3. It accepts the word *aba* since after reading *a*, it is in state s_1. After reading *b*, it is still in state s_1. After reading the second *a*, it is in state s_3, which is an acceptance state. One can see that it also accepts *abbba* and *bb*, so they are in the language accepted by the given automaton. However *bbb*, *abab*, and *abb* are not. Notice that any string beginning with two *a*s or two *b*s is accepted only if the string is not extended. Also, if three *a*s occur in the string, the string is not accepted. The state s_4 is an example of a **sink state**. Once the automaton is in the sink state, it can never leave this state again, regardless of the letter read.

Since Υ is a function, a deterministic automaton can always read the entire string. We shall later define a nondeterministic automaton which may not always be able to read the entire string. In such a case the word cannot be accepted.

Example 3.1   Consider the automaton with state diagram

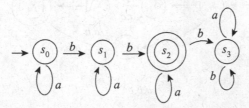

having $\Sigma = \{a, b\}$, starting state s_0, and acceptance states s_0, s_1, and s_2. It obviously accepts the word *bb*. In each state, there is a loop for *a* so that if a is read then the state does not change. This enables us to read as many as as desired without changing states, before reading another b. Thus the automaton reads *aababaaa*, *baabab*, *baaab*, *babaaa*, *aabaabaa*, and in fact we can read any word in the language described by the regular expression $(a^*ba^*ba^*) \vee (a^*ba^*) \vee a^*$. This language can also be described as the set of all words containing at most two bs. Notice that s_3 is a sink state.

Example 3.2   Consider the automaton with state diagram

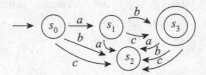

which we simplify as

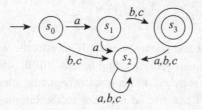

to decrease the number of arrows. This automaton obviously accepts only the words *ab* and *ac*. This language may be described by the regular expression $a(b \vee c)$. Notice that the sink state s_2 eliminates all other words from the language.

Example 3.3   Consider the automaton with state diagram

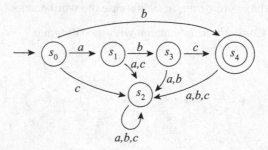

The only words accepted are *b* and *abc*. Therefore the expression for the language accepted is $b \vee abc$.

Example 3.4 Consider the automaton with state diagram

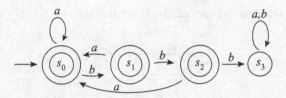

In this automaton, if three consecutive bs are read, then the automaton is in state s_3, which is a sink state and is not an acceptance state. This is the only way to get to s_3 and every other state is an acceptance state. Thus the language accepted by this automaton consists of all words which do not have three consecutive bs. An expression for this language is

$$(\mathbf{a} \vee (\mathbf{ba}) \vee (\mathbf{bba}))^*(\lambda \vee \mathbf{b} \vee (\mathbf{bb})).$$

As previously mentioned, the automata that we have been discussing are called **deterministic automata** since in every state and for every value of the alphabet that is read, there is one and only one state in which the automata can be. In other words, $\Upsilon : Q \times \Sigma \to Q$ is a function. It is often convenient to relax the rules so that Υ is no longer a function, but a relation. If we again consider Υ as a set of rules, given $a \in \Sigma$ and $s \in Q$, the rules may allow advancement to each of several states or there may not be a rule which does not allow it to go to any state after reading a in state s. In the latter case, the automaton is "hung up" and can proceed no further. This cannot occur with a deterministic automaton.

Although the definition of a nondeterministic automaton varies, we shall use the following definition:

Definition 3.2 *A **nondeterministic automaton**, denoted by*

$$(\Sigma, Q, s_0, \Upsilon, F)$$

consists of a finite alphabet Σ, a finite set Q of states, and a function

$$\Upsilon : Q \times \Sigma \to \mathcal{P}(Q)$$

*called the **transition function**. The set Q contains an element s_0 and a subset F containing one or more acceptance states. (Note that $\mathcal{P}(Q)$ is the power set of Q.)*

Thus given $a \in \Sigma$ and $s \in Q$, there may be a-arrows from s to several different states or to no state at all. By definition, a deterministic automaton is also considered to be a nondeterministic automaton. A nondeterministic automaton often simplifies the state diagram and eliminates the need for a sink state. In Example 3.2, the state diagram can be simplified to

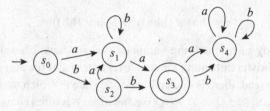

Note that in reading *aa*, after reading the first *a*, the automaton is in state s_3, and when the second *a* is read the automaton "hangs up", since there is no *a* arrow out of state s_3.

Example 3.5   The deterministic automaton represented by

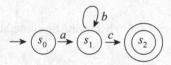

can be simplified using a nondeterministic automaton by simply eliminating state s_4 and all arrows into or out of this state.

Example 3.6   It is easily seen that the automaton with state diagram

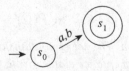

accepts the language with regular expression ***ab**c***.

Example 3.7   The automaton with state diagram

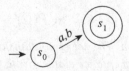

accepts the language with regular expression ***a* ∨ *b***.

Example 3.8   The automaton with state diagram

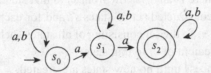

accepts the language with regular expression aa^*bb^*.

Example 3.9   The automaton with state diagram

accepts the language consisting of strings with at least two as and so may be written as $(a \lor b)^*a(a \lor b)^*a(a \lor b)^*$.

Obviously any language accepted by a deterministic automaton is accepted by a nondeterministic automaton since the set of deterministic automata is a subset of the set of nondeterministic automata. In the following theorem, however, we shall see that any language accepted by a nondeterministic automaton is also accepted by a deterministic automaton.

Theorem 3.1   *For each nondeterministic automaton, there is an equivalent deterministic automaton that accepts the same language.*

We demonstrate how to construct a deterministic automaton which accepts the language accepted by a nondeterministic automaton. We shall later give a formal proof that a language is accepted by a deterministic automaton if and only if it is accepted by a nondeterministic automaton. If Q is the set of states for the nondeterministic automaton, we shall use elements of $\mathcal{P}(Q)$, i.e. the set of subsets of Q, as states for the deterministic automaton which we are constructing. Some of these states may not be used since they do not occur on any path which leads to acceptance state. Hence they could be removed and greatly simplify the deterministic automaton created. However, for our purpose, we are only interested in showing that a deterministic automaton can be created.

In general we have the following procedure for constructing a deterministic automaton

$$M = (\Sigma, Q', \{s_0\}, \Upsilon', F')$$

from a nondeterministic automaton.

$$N = (\Sigma, Q, s_0, \Upsilon, F).$$

(1) Begin with the state $\{s_0\}$ where s_0 is the start state of the nondeterministic automaton.
(2) For each $a_i \in \Sigma$, construct an a_i arrow from $\{s_0\}$ to the set consisting of all states such that there is an a_i-arrow from s_0 to that state.
(3) For each newly constructed set of states s_j and for each $a_i \in \Sigma$ construct an a_i arrow from s_j to the set consisting of all states such that there is an a_i arrow from an element of s_j to that state.
(4) Continue this process until no new states are created.
(5) Make each set of states s_j, that contains an element of the acceptance set of the nondeterministic automaton, into an acceptance state.

Example 3.10   Consider the nondeterministic automaton N

Construct an a-arrow from $\{s_0\}$ to the set of all states so that there is an a-arrow from s_0 to that state. Since there is an a-arrow from s_0 to s_0 and an a-arrow from s_0 to s_1, we construct an a-arrow from $\{s_0\}$ to $\{s_0, s_1\}$. There is no b-arrow from s_0 to any state. Hence the set of all states such that there is a b-arrow to one of these states is empty and we construct a b-arrow from $\{s_0\}$ to the empty set $\varnothing$. We now consider the state $\{s_0, s_1\}$. We construct an a-arrow from $\{s_0, s_1\}$ to the set of all states such that there is an a-arrow from either s_0 or s_1 to that state. Thus we construct an a-arrow from $\{s_0, s_1\}$ to itself. We construct a b-arrow from $\{s_0, s_1\}$ to the set of all states such that there is a b-arrow from either s_0 or s_1 to that state. Thus construct a b-arrow from $\{s_0, s_1\}$ to $\{s_2\}$. Since there are no a-arrows or b-arrows from any state in the empty set to any other state, we construct an a-arrow and a b-arrow from the empty set to itself. Consider $\{s_2\}$. Since there is no a-arrow from s_2 to any other state, we construct an a-arrow from $\{s_2\}$ to the empty set. Since the only b-arrow from s_2 is to itself, we construct a b-arrow from $\{s_2\}$ to itself. The acceptance states consist of all sets which contain an element of the terminal set of N. In this case $\{s_2\}$ is the only acceptance state. We have now completed the state diagram

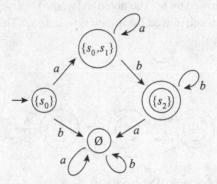

which is easily seen to be the state diagram of a deterministic automaton. This automaton also reads the same language as N, namely the language described by the expression aa^*bb^*.

Example 3.11 Given the nondeterministic automaton

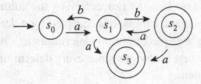

using the same method as above we complete the deterministic automaton

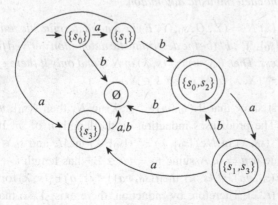

At this point we introduce a new notation. The ordered pair (s_i, w) indicates that the automaton is in state s_i and still has input w left to read. For example,

$(s_2, abbb)$ indicates that the automaton is in state s_2 and must still read $abbb$. Assume that we have (s_i, aw) $w \in \Sigma^+$. Thus the automaton is in state s_i and must still read a followed by w. The notation $(s_i, aw) \vdash (s_j, w)$ means that the automaton has read a and moved from state s_i to state s_j. Therefore $\Upsilon(s_i, a) = s_j$. In the automaton

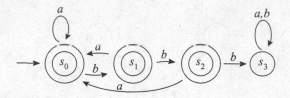

we have $(s_2, bab) \vdash (s_3, ab)$. We also have

$$(s_0, babba) \vdash (s_1, abba) \vdash (s_0, bba) \vdash (s_1, ba) \vdash (s_2, a) \vdash (s_0, \lambda).$$

If we have $(s_i, w_i) \vdash (s_j, w_j) \vdash \cdots \vdash (s_m, w_m)$, we denote this by $(s_i, w_i) \vdash^* (s_m, w_m)$. We also let $(s, w) \vdash^* (s, w)$. Thus a word w is accepted by an automaton if and only if $(s_0, w) \vdash^* (s, \lambda)$ where s is an acceptance state. In our example $(s_0, bababb) \vdash^* (s_0, \lambda)$, so $bababb$ is accepted by the automaton.

We shall now prove that a language is accepted by a deterministic automaton if and only if it is accepted by a nondeterministic automaton. We begin with two lemmas. The first is obvious since every deterministic automaton is a nondeterministic automaton.

Lemma 3.1 *Every language accepted by a deterministic automaton is accepted by a nondeterministic automaton.*

Lemma 3.2 *Let $N = (\Sigma, Q, s_0, \Upsilon, F)$ be a nondeterministic automaton and $M = (\Sigma, Q', \{s_0\}, \Upsilon', F')$ be the deterministic automaton derived from N using the above process. Then $(s_0, w) \vdash^* (s, \lambda)$ in N if and only if there exists X such that $(\{s_0\}, w) \vdash^* (X, \lambda)$ in M where $s \in X$.*

Proof We first show that if $(s_0, w) \vdash^* (s, \lambda)$ in N, then $(\{s_0\}, w) \vdash^* (X, \lambda)$ where $s \in X$. The proof uses induction on the length n of w. If $n = 0$, we have $(s_0, \lambda) \vdash^* (s_0, \lambda)$ in N, $(\{s_0\}, \lambda) \vdash^* (\{s_0\}, \lambda)$ in M, and $s_0 \in \{s_0\}$, so the statement is true if $n = 0$. Assume $w = va \in \Sigma^+$ has length $k + 1$, so v has length n. Since $(s_0, va) \vdash^* (s, \lambda)$, then $(s_0, va) \vdash^* (t, a) \vdash (s, \lambda)$ for some $t \in Q$ and $(s_0, v) \vdash^* (t, \lambda)$. Therefore by induction, there exist Y so that $t \in Y$ and $(\{s_0\}, v) \vdash^* (Y, \lambda)$. Since $t \in Y$ and $(t, a) \vdash (s, \lambda)$ in N, $(Y, a) \vdash (X, \lambda)$ for some X where $s \in X$. Therefore $(\{s_0\}, va) \vdash^* (Y, a) \vdash (X, \lambda)$ or $(\{s_0\}, va) \vdash^* (X, \lambda)$ where $s \in X$.

Conversely, we show that if $(\{s_0\}, w) \vdash^* (X, \lambda)$ in M, then $(s_0, w) \vdash^* (s, \lambda)$ in N where $s \in X$. We again use induction on n, the length of the word w. Assume there exists X such that $(\{s_0\}, w) \vdash^* (X, \lambda)$ in M where $s \in X$. If $n = 0$, we have $(\{s_0\}, \lambda) \vdash^* (\{s_0\}, \lambda)$ in M, $(s_0, \lambda) \vdash^* (s_0, \lambda)$ in N, and $s_0 \in \{s_0\}$, so the statement is true if $n = 0$. Given $(\{s_0\}, va)$ with length $k + 1$, so v has length n. Assume $(\{s_0\}, va) \vdash^* (Y, a) \vdash (X, \lambda)$. Therefore $(\{s_0\}, v) \vdash^* (Y, \lambda)$. By induction, $(s_0, v) \vdash^* (t, \lambda)$ for all t in Y and hence $(s_0, va) \vdash^* (t, a)$ for all t in Y. By definition, since $(Y, a) \vdash (X, \lambda)$ and $(t, a) \vdash (s, \lambda)$, then $(s_0, w) \vdash^* (s, \lambda)$ in N for $s \in X$. $\qquad\square$

We are now able to prove the desired Theorem 3.1.

Theorem 3.2 *A language is accepted by a deterministic automaton if and only if it is accepted by a nondeterministic automaton.*

Proof To show this we need only show that a word is accepted by a nondeterministic automaton if and only if it is accepted by the corresponding deterministic automaton. If $(s_0, w) \vdash^* (s, \lambda)$ where s is an acceptance state in the nondeterministic automaton, then $(\{s_0\}, w) \vdash^* (X, \lambda)$ where X contains an acceptance state. Hence X is an acceptance state. Assume X is an acceptance state, then it contains an acceptance state r from the nondeterministic automaton. But by the previous lemma, if $(\{s_0\}, w) \vdash^* (X, \lambda)$ and $r \in X$ then $(s_0, w) \vdash^* (r, \lambda)$. Therefore r is an acceptance state. $\qquad\square$

At this point we shall define an extended nondeterministic automaton and prove that a language is accepted by an extended nondeterministic automaton if and only if it is accepted by a nondeterministic automaton (and hence a deterministic automaton).

Using a nondeterministic automaton, we can extend the automaton so that $(\Sigma^+, Q, s_0, \Upsilon, F)$ consists of Σ^+, a finite set Q of states, and a function $\Upsilon : \Sigma^+ \times Q \to \mathcal{P}(Q)$, called the transition function. Thus Υ reads words instead of letters. This can be changed back to reading letters by adding new nonterminal states. If Υ reads the word $w = a_1 a_2 \cdots a_k$, and moves from state s to state s', add states $\sigma_2 \sigma_3 \cdots \sigma_k$, and let $\Upsilon(s, a_1) = \sigma_2$, $\Upsilon(\sigma_2, a_2) = \sigma_3$, $\Upsilon(\sigma_3, a_3) = \sigma_4, \ldots, \Upsilon(\sigma_{k-1}, a_{k-1}) = \sigma_k$, and $\Upsilon(\sigma_k, a_k) = s'$. This forms a nondeterministic automata, but we can form a deterministic automata with the same language as shown above.

If we allow the automaton to pass from one state s_i to another state s_j without reading a letter of the alphabet, this may be shown as the automaton having an edge from s_i to s_j with label λ. Thus paths may contain one or more λ's. Such an automaton is said to have λ-moves. We can then have an automaton with the form $(\Sigma^*, Q, s_0, \Upsilon, F)$.

Formally a finite automaton $M = (\Sigma, Q, s_0, \Upsilon, F)$ with λ-moves has the property that Υ maps $Q \cup \{\lambda\}$ to Q. We wish to create a deterministic automata $M' = (\Sigma, Q', s_0', \Upsilon', F')$ containing no λ-moves with the same language. Thus $M(L) = M'(L)$. Given a letter q in Σ, define $E(q)$ to be all the states that are reachable from q without reading a letter in the alphabet. Thus $E(q) = \{p : (q, w) \vdash (p, w)$. In our construction, the set of states of M' is a subset of $\mathcal{P}(Q)$. The state $s_0' = E(s_0)$, and F' is a set containing an element of F. For each element a of Σ, define Υ' by $\Upsilon'(P, a) = \bigcup_{p \in P} E(\Upsilon(p, a))$.

We first show that M' is deterministic. It is certainly single valued. Further $\Upsilon'(P, a)$ will always have a value even if it is the empty set.

We must now show that $M(L) = M'(L)$. To do this we show that for any states p and q in Q, and any word w in Σ^*

$$(p, w) \vdash^* (q, \lambda) \text{ in } M \text{ if and only if } (E(p), w) \vdash^* (P, \lambda) \text{ in } M'$$

for some P containing q. From this it will follow that

$$(s_0, w) \vdash^* (f, \lambda) \text{ in } M \text{ if and only if } (E(s_0), w) \vdash^* (P, \lambda) \text{ in } M'$$

for some P containing f, where $f \in F$.

We prove this using induction of the length of w. If $|w| = 0$, then $w = \lambda$, and it must be shown that

$$(p, \lambda) \vdash^* (q, \lambda) \text{ in } M \text{ if and only if } (E(p), \lambda) \vdash^* (P, \lambda) \text{ in } M'$$

for some P containing q. Now $(p, \lambda) \vdash^* (q, \lambda)$ if and only if $q \in E(p)$; but since M' is deterministic and no letter is read, then $P = E(p)$ and $p \in E(p)$. Therefore the statement is true if $|w| = 0$.

Assume the statement is true for all strings having nonnegative length k. We now have to prove the statement is true for any string w with length $k + 1$.

$\Rightarrow$: Assume $w = va$ for some letter a and w and $(p, w) \vdash^* (q, \lambda)$ so that

$$(p, va) \vdash^* (q_1, a) \vdash (q_2, \lambda) \vdash^* (q, \lambda)$$

where at the end, possibly no letters of the alphabet are read. Since $(p, va) \vdash^* (q_1, a)$ then $(p, v) \vdash^* (q_1, \lambda)$ and, by induction, $(E(p), v) \vdash^* (R, \lambda)$ for some R containing q_1. But since $(q_1, a) \vdash (q_2, \lambda)$, by construction, $E(q_2) \subseteq \Upsilon'(R, a)$, and since $(q_2, \lambda) \vdash^* (q, \lambda)$, $q \in E(q_2)$ by definition of E, and hence $q \in \Upsilon'(R, a)$. Therefore $(R, a) \vdash ((P, \lambda)$ for some P containing q by definition of Υ' and $(E(p), va) \vdash^* (R, a) \vdash ((P, \lambda)$ for some P containing q.

In M', assume $(E(p, va)) \vdash^* (R, a) \vdash (P, \lambda)$ where $q \in P$ and $\Upsilon'(R, a) = P$. By definition $\Upsilon'(R, a) = \bigcup_{r \in R} E(\Upsilon(r, a))$. There exists some state $r \in R$ such that $\Upsilon(r, a) = s$ and $q \in E(s)$. Therefore $(s, \lambda) \vdash^* (q, \lambda)$ by definition of $E(s)$. By the induction hypotheses $(p, v) \vdash^* (r, \lambda)$. Therefore $(p, va) \vdash^*$ $(r, a) \vdash (s, \lambda) \vdash^* (q, \lambda)$.

Example 3.12 Given the automaton $(M = (\Sigma, Q, s_0, \Upsilon, F)$

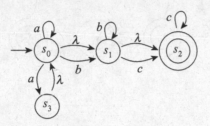

which has λ-moves, we construct $M' = (\Sigma, Q', s_0', \Upsilon', F')$ containing no λ-moves: $E(s_0) = \{s_0, s_1, s_2\}$, $E(s_1) = \{s_1, s_2\}$, $E(s_2) = \{s_2\}$, and $E(s_3) = \{s_0, s_1, s_2, s_3\}$. Denote these sets by s_0', s_1', s_2', and s_3' respectively. Then Υ' is given by the following table

	a	b	c
s_0'	s_3'	s_1'	s_2'
s_1'	$\emptyset$	s_1'	s_2'
s_2'	$\emptyset$	$\emptyset$	s_2'
s_3'	s_3'	s_1'	s_2'

giving the λ- free automaton

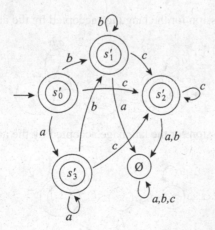

Both automata generate the language $a^*b^*c^*$.

Exercises

(1) Which of the following words are accepted by the automaton?

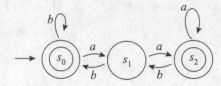

    (a) *abba.*
    (b) *aabbb.*
    (c) *babab.*
    (d) *aaabbb.*
    (e) *bbaab.*

(2) Which of the following words are accepted by the automaton?

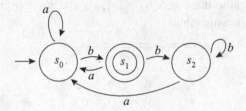

    (a) *aaabb.*
    (b) *abbbabbb.*
    (c) *bababa.*
    (d) *aaabab.*
    (e) *bbbabab.*

(3) Write an expression for the language accepted by the automaton

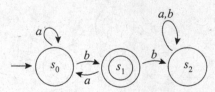

(4) Write an expression for the language accepted by the automaton

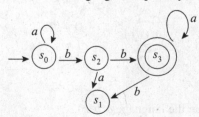

(5) Write an expression for the language accepted by the automaton

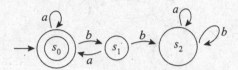

(6) Write an expression for the language accepted by the automaton

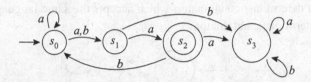

(7) Find a deterministic automaton which accepts the language expressed by **aa*bb*cc***.

(8) Find a deterministic automaton which accepts the language expressed by **(a*ba*ba*b)***.

(9) Find a deterministic automaton which accepts the language expressed by **(a*(ba)*bb*a)***.

(10) Find a deterministic automaton which accepts the language expressed by **(a*b) ∨ (b*a)***.

(11) Find a nondeterministic automaton which accepts the language expressed by **aa*bb*cc***.

(12) Find a nondeterministic automaton which accepts the language expressed by **(a*b) ∨ (c*b) ∨ (ac)***.

(13) Find a nondeterministic automaton which accepts the language expressed by **(a ∨ b)*(aa ∨ bb)(a ∨ b)***.

(14) Find a nondeterministic automaton which accepts the language expressed by **((aa*b) ∨ bb*a)ac***.

(15) Find a deterministic automaton which accepts the same language as the nondeterministic automaton

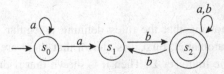

(16) Find a deterministic automaton which accepts the same language as the nondeterministic automaton

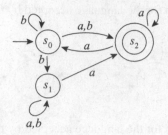

(17) Find a deterministic automaton which accepts the same language as the nondeterministic automaton

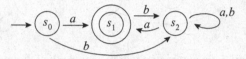

(18) Find a deterministic automaton which accepts the same language as the nondeterministic automaton

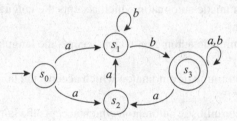

3.2 Kleene's Theorem

In this section we show Kleene's Theorem which may be stated as follows:

Theorem 3.3 *A language is regular if and only if it is accepted by an automaton.*

We begin by showing that the rules defining a regular language can be duplicated by an automaton. First it is shown that there are automata which accept subsets of the finite set Σ. Then it is shown that if there are automata that accept languages L_1 and L_2, we can construct automata that accept $L_1 L_2$, L_1^*, and $L_1 \cup L_2$. Thus we show that for every regular language over a finite set Σ, there is a nondeterministic automaton that accepts that language.

The set is λ; it is accepted by the automaton with state diagram.

It may be preferable to create a state s_1, which is not an acceptance state, and have all arrows from both s_0 and s_1 go to s_1.

We next show that for every finite subset of Σ, there is an automaton that reads that subset. If the automaton has no acceptance state, then the language accepted is the empty set. The elements of the set $\{a_1, a_2, a_3, \ldots, a_n\}$ are accepted by the automaton with state diagram

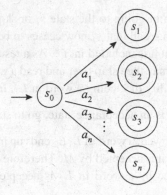

In particular if $a \in \Sigma$, it is accepted by the automaton with state diagram

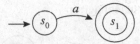

We next show that if regular languages with expression L_1 and L_2 are both accepted by automata, then their concatenation $L_1 L_2$ is also accepted by an automaton. Assume L_1 is accepted by the automaton $M_1 = (\Sigma, Q, s_0, \Upsilon, F)$, and L_2 is accepted by the automaton $M_2 = (\Sigma, Q', s_0', \Upsilon', F')$. We shall consider both automata to be deterministic without loss of generality. We now define a new automaton $M = (\Sigma, Q'', s_0'', \Upsilon'', F'')$ which is essentially the first automaton followed by the second automaton. Put simply, place the state diagram for M_2 after the state diagram for M_1. If, for $a \in \Sigma$, there is an a-arrow from any state s in the state diagram for M_1 to an acceptance state in the state diagram for M_1, then change the acceptance state into a nonacceptance state and also place an a-arrow from s to the starting state in the state diagram for M_2. This is the state diagram for M. Thus the set of states $Q'' = Q \cup Q'$, so that Q'' consists of all the states in M_1 and M_2. We shall assume that M_1 and M_2

have no common states. If they do, we can always relabel them. Since we want to begin in M_1, we let s_0 be the starting state of M so that $s_0'' = s_0$. Since we want to finish in M_2, we let the set of acceptance states be F' so that $F'' = F'$. We define the rules for Υ'' as follows.

If the rule

> If in state s_i and a is read, go to state s_j

is in Υ and s_j is not an acceptance state then include this rule in Υ''. If s_j is an acceptance state then include this rule in Υ'' and also include the rule

> If in state s_i and a is read, go to state s_0'.

Hence there is the option of going to the state s_j or skipping over to s_0' in the second automaton. Again recall that s_j now ceases to be an acceptance state.

If the rule is in Υ' then it is included in Υ''. As a result, if the automaton M has read a word in L_1, it may then skip over and read a word in L_2. As a special case, consider the possibility of λ being a word in L_1. Include the rule

> If state s_0 is an acceptance state, go to state s_0'.

When these rules are followed, a word in L_1L_2 ends up in an acceptance state of M_2 and hence Υ'' so that it is accepted by M. Therefore every string consisting of a word in L_1 followed by a word in L_2 is accepted by M, and L_1L_2 is accepted by M.

Example 3.13 Let L_1 be the language described by the language $(ab)^*c$ and having automaton M_1 with state diagram

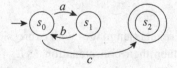

Let L_2 be the language described by the language ab^*c^* and having automaton M_2 with state diagram

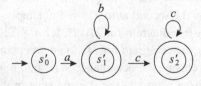

To find the state diagram for the language L_1L_2, place the state diagram for M_2 after the state diagram for M_1. Since there is a c-arrow from s_0 to s_2, and s_2 is

an acceptance state, add a c-arrow from s_0 to s_0'. The state diagram

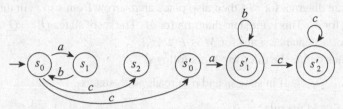

is the state diagram for M, the automaton for $L_1 L_2$.

Example 3.14 Let L_1 and L_2 and their respective automata be the same as those in the previous example. To find the automaton for the language for $L_2 L_1$ is slightly more complicated. First we place the state diagram for M_1 after the state diagram for M_2. There is an a-arrow from s_0' to s_1' and s_1' is an acceptance state, so place an a-arrow from s_0' to s_0. There is a b-arrow from s_1' to s_1' and s_1' is an acceptance state, so place a b-arrow from s_1' to s_0. There is a c-arrow from s_1' to s_2' and s_2' is an acceptance state, so place a c-arrow from s_1' to s_0. There is a c-arrow from s_2' to s_2' and s_2' is an acceptance state, so place a c-arrow from s_2' to s_0. Then change s_1', s_2' so that they are not acceptance states. Thus we have the state diagram

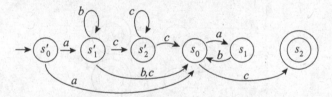

which is the state diagram for M, the automaton for $L_2 L_1$.

Similarly we show that if L is a language accepted by an automaton $M_1 = (\Sigma, Q, s_0, \Upsilon, F)$ then L^* is also accepted by an automaton. We now define a new automaton $M = (\Sigma, Q', s_0', \Upsilon', F')$ which is essentially the same as M_1 except M is looped to itself. Let M be defined as follows: Create a new state s_0', and make it an acceptance state. We include state s_0' so M will accept the empty word. For each rule

If in state s_0 and a is read, go to state s_j

for $a \in \Sigma$, add the rule

If in state s_0' and a is read, go to state s_j.

Thus if there is an a-arrow from s_0 to s_j, there is an a-arrow from s_0' to s_j. Include all of the current rules for M_1 in M. In addition, for $a \in \Sigma$, if there is an a-arrow from any state s in the state diagram for M_1 to an acceptance state in the state diagram for M_1, then also place an a-arrow from s to s_0 in the state diagram for M. This is the state diagram for M. The set of states $Q' = Q \cup \{s_0'\}$. The set of acceptance states for M is $F \cup \{s_0'\}$.

We thus define the rules for Υ' as follows:

> If in state s_0 and a is read, go to state s_j

for $a \in \Sigma$, add the rule

> If in state s_0' and a is read, go to state s_j.

If the rule is in Υ, then include this rule in Υ'. If s_j is an acceptance state and

> If in state s_i and a is read, go to state s_j

then also include the rule

> If in state s_i and a is read, go to state s_0.

Hence there is the option of going to the acceptance state s_j or skipping over to s_0.

Example 3.15 Let the diagram

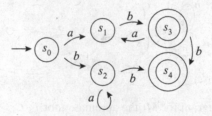

be the state diagram for the automaton accepting the language L, then the diagram

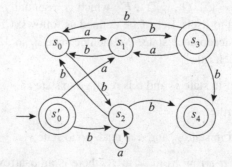

is the state diagram for the automaton accepting the language L^*.

Since the relationship between an automaton and its state diagram should be evident by now, in the future we will identify an automaton with its state diagram. Given automata M and M' accepting languages L and L' respectively, we now wish to construct an automaton M'' which accepts $L \cup L'$. We wish to read a word simultaneously in M and M', and accept it if it is accepted by either M or M'.

We now show how to construct M'' which accepts $L \cup L'$ given automata $M = (\Sigma, Q, s_0, \Upsilon, F)$ and $M' = (\Sigma, Q', s_0', \Upsilon', F')$ If s_0 and s_0' are the initial states of M and M' respectively then construct a new initial state s_0'', which is an acceptance state if either s_0 or s_0' is, and let $M'' = (\Sigma, Q'', s_0'', \Upsilon'', F'')$ where $Q'' = Q \cup Q' \cup \{s_0''\}$ and $F'' = F \cup F'$. Let $\Upsilon'' = \Upsilon \cup \Upsilon'$ together with the following rules: If there is a rule

> If in state s_0 and a is read, go to state s_j

for $a \in \Sigma$ in Υ include the rule

> If in state s_0'' and a is read, go to state s_j

in Υ''.

If there is a rule

> If in state s_0' and a is read, go to state s_j'

for $a \in \Sigma$ in Υ' include the rule

> If in state s_0'' and a is read, go to state s_j'

in Υ''.

Example 3.16 Let M be the automaton

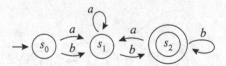

and M' be the automaton

Using the above procedure we have the automaton M'' which accepts the union of $M(L)$ and $M'(L)$ given by

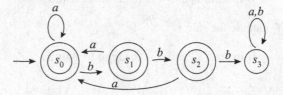

Example 3.17 Let M be the automaton

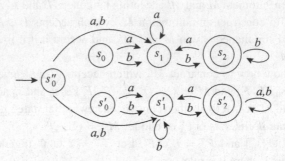

and M' be the automaton

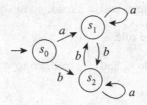

Using the above procedure we have the automaton M which accepts the union of $M_1(L)$ and $M_1'(L)$

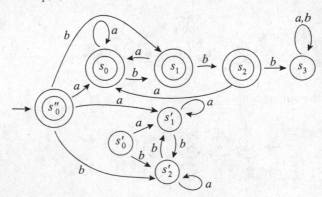

An alternative method for finding the union of two automata is now given. If Q is the set of states for M and Q' is the set of states for M', let $Q \times Q' = \{(s_i, s'_j) : s_i \in Q, s'_j \in Q'\}$ be the set of states for M. If there is an a_i-arrow from s_i to s_k in M and an a_i-arrow from s'_l to s'_m in M', then construct an a_i-arrow from (s_i, s'_l) to (s_k, s'_m). In this way, we read the same letter simultaneously in M and M'. Since a word is accepted if it is accepted by either M or M', if s_i is a terminal state in M or s'_j is a terminal state in M', then we want (s_i, s'_j) to be a terminal state in M''. Therefore if F is the set of terminal states of M and F' is the set of terminal states of M', the set of terminal states for M'' is $Q \times F' \cup F \times Q'$. We require that both M and M' be deterministic since we do not want M'' to "hang" in one automaton before being accepted in the other. This is no restriction since we have shown that any language accepted by a nondeterministic automaton is also accepted by a deterministic automaton.

Example 3.18 Let M be the automaton

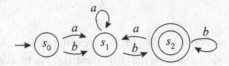

and M' be the automaton

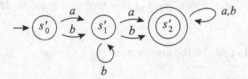

It may be that all of the states in $Q_1 \times Q'_1$ are not needed. We begin with (s_0, s'_0) as the start state. Since there is an a-arrow from s_0 to s_1 and an a-arrow from s'_0 to s'_1, we construct an a-arrow from (s_0, s'_0) to (s_1, s'_1). Since there is also a b-arrow from s_0 to s_1 and a b-arrow from s'_0 to s'_1, we construct a b-arrow from (s_0, s'_0) to (s_1, s'_1). Since there is an a-arrow from s_1 to s_1 and an a-arrow from s'_1 to s'_2, we construct an a-arrow from (s_1, s'_1) to (s_1, s'_2). There is a b-arrow from s_1 to s_2 and a b-arrow from s'_1 to s'_2, so we construct a b-arrow from (s_1, s'_1) to (s_2, s'_2). Continuing at (s_1, s'_2), there is an a-arrow from s_1 to s_1 and an a-arrow from s'_2 to s'_2, we construct an a-arrow from (s_1, s'_2) to (s_1, s'_2). There is a b-arrow from s_1 to s_2 and a b-arrow from s'_2 to s'_2, so we construct a b-arrow from (s_1, s'_2) to (s_2, s'_2). Continuing at (s_2, s'_1), there is an a-arrow from s_2 to s_1 and an a-arrow from s'_1 to s'_2, so we construct an a-arrow from (s_2, s'_1) to (s_1, s'_2).

There is a b-arrow from s_2 to s_2 and a b-arrow from s_1' to s_2', so we construct a b-arrow from (s_2, s_1') to (s_2, s_2'). Finally, consider state (s_2, s_2'). There is an a-arrow from s_2 to s_1 and an a-arrow from s_2' to s_2', so we construct an a-arrow from (s_2, s_2') to (s_1, s_2'). There is a b-arrow from s_2 to s_2 and a b-arrow from s_2' to s_2', so we construct a b-arrow from (s_2, s_2') to (s_2, s_2'). The terminal states are (s_1, s_2'), (s_2, s_1'), and (s_2, s_2'). Thus M'' is the automaton

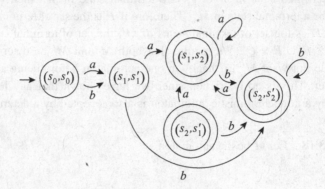

Note that $aabb$ is accepted by M. In M'', reading $aabb$ takes us from state (s_0, s_0') to state (s_1, s_1') to state (s_1, s_2') to state (s_2, s_2') to state (s_2, s_2'). Since (s_2, s_2') is a final state in M'', M'' accepts $aabb$. Note also that aba is accepted by M'. In M'', reading aba takes us from state (s_0, s_0') to state (s_1, s_1') to state (s_2, s_1') to state (s_1, s_2'). Since (s_1, s_2') is a terminal state in M'', M'' accepts aba.

Example 3.19 Let M_1 be the automaton

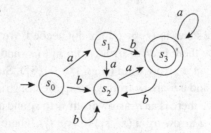

and M_1' be the automaton

Using the same process as in the previous example, we find that M is the automaton

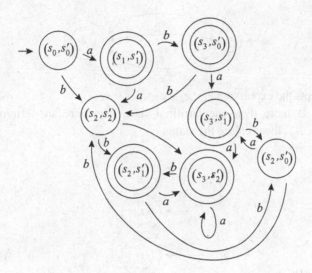

We have now shown that every operation in a regular language can be duplicated by an automaton. Hence we have the following lemma.

Lemma 3.3 *For every regular language L, there exists an automaton M so that L is the language accepted by M.*

The formal proof that a language accepted by an automaton is regular gives no procedure for actually converting an automaton to a language accepted by an automaton. Before giving a proof that the language accepted by an automaton M is regular, we first give some examples where, given a certain automaton, we can construct the language accepted by this automaton. This is not part of the proof but only an illustration. To perform this construction, we use transition graphs. These are merely finite state machines which read strings of a regular expression rather than elements of Σ to change states. One of the regular expressions we shall use is the empty word λ so that one may change states reading the empty word which is equivalent to changing states without reading anything. The form of the transition graph will become obvious as we use them.

The process for constructing the regular expression is to first have only one initial and one terminal state. We then eliminate one state at a time from the state diagram and resulting transition graphs and in each case get a transition graph with **e**-arrows between states, where **e** is a regular expression. Eventually

we get a transition graph of the form

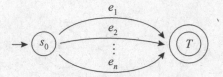

which accepts the expression $\mathbf{e}_1 \vee \mathbf{e}_2 \vee \mathbf{e}_3 \vee \cdots \vee \mathbf{e}_n$.

If there is more than one terminal state, say there are terminal states $t_1, t_2, t_3, \ldots, t_m$, then replace the states

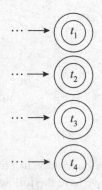

with

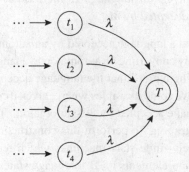

Note that this new diagram accepts the same language as M.

To eliminate the state s_i we use the following rules.

(1) If the diagram

occurs, replace it with

$$\xrightarrow{\quad} \enspace s_i \enspace \circlearrowright^{(a \vee b \vee c)^*} \xrightarrow{\quad}$$

More generally if the diagram

$$\xrightarrow{\quad} \enspace s_i \enspace \circlearrowright^{e_1, e_2, \dots e_k} \xrightarrow{\quad}$$

occurs, where $e_1, e_2, e_3, \cdots, e_k$ are regular expressions, then replace it with the diagram

$$\xrightarrow{\quad} \enspace s_i \enspace \circlearrowright^{(e_1 \vee e_2 \vee e_3 \vee \dots \vee e_k)} \xrightarrow{\quad}$$

(2) If the diagram

$$\xrightarrow{\quad} s_{i-1} \xrightarrow{\ a\ } \enspace s_i \enspace \circlearrowright^{b} \xrightarrow{\ c\ } s_{i+1}$$

occurs, then replace it with the diagram

$$\xrightarrow{\quad} s_{i-1} \xrightarrow{\ ab^*c\ } s_{i+1}$$

More generally if the diagram

$$\xrightarrow{\quad} s_{i-1} \xrightarrow{\ e_1\ } \enspace s_i \enspace \circlearrowright^{e_2} \xrightarrow{\ e_3\ } \enspace s_{i+1} \enspace \xrightarrow{\quad}$$

occurs, where e_1, e_2, e_3 are regular expressions, then replace it with the diagram

$$\xrightarrow{\quad} s_{i-1} \xrightarrow{\ e_1 e_2^* e_3\ } \enspace s_{i+1} \enspace \xrightarrow{\quad}$$

In particular, when $e_2 = \lambda$, then $e_1 e_2^* e_3$ becomes $e_1 e_3$ so that the diagram

$$\xrightarrow{\quad} \enspace s_{i-1} \enspace \xrightarrow{\ a\ } \enspace s_i \enspace \xrightarrow{\ b\ } \enspace s_{i+1}$$

is replaced by the diagram

$$\longrightarrow (s_{i-1}) \xrightarrow{\ ab\ } (s_{i+1})$$

(3) If the diagram

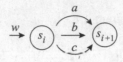

occurs, then replace it with the diagram

$$\xrightarrow{w} (s_i) \xrightarrow{\ a \vee b \vee c\ } (s_{i+1})$$

More generally if the diagram

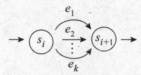

occurs, where $e_1, e_2, e_3, \cdots, e_k$ are regular expressions, then replace it with the diagram

$$\longrightarrow (s_i) \xrightarrow{\ e_1 \vee e_2 \vee \ldots \vee e_k\ } (s_{i+1}) \longrightarrow$$

(4) If the diagram

$$\longrightarrow (s_{i-1}) \underset{b}{\overset{a}{\rightrightarrows}} (s_2) \xrightarrow{\ c\ } (s_{i+1})$$

occurs, then replace it with the diagram

$$\longrightarrow (s_{i-1}) \xrightarrow{\ a(ba)^*c\ } (s_{i+1})$$

More generally if the diagram

$$\longrightarrow (s_{i-1}) \overset{e_1}{\underset{e_2}{\rightleftarrows}} (s_2) \xrightarrow{\ e_3\ } (s_{i+1})$$

occurs, where $\mathbf{e}_1, \mathbf{e}_2, \mathbf{e}_3$ are regular expressions, then replace it with the diagram

$$\rightarrow \boxed{s_{i-1}} \xrightarrow{e_1(e_2e_1)^*e_3} \boxed{s_{i+1}}$$

(5) If the diagram

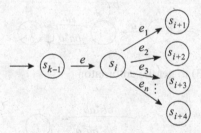

occurs, then replace it with the diagram

$$\longrightarrow \boxed{s_{k-1}} \overset{ab}{\underset{ac}{\rightrightarrows}} \begin{matrix} \boxed{s_{i+1}} \\ \boxed{s_{i+2}} \end{matrix}$$

More generally if the diagram

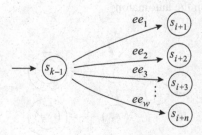

occurs, where $\mathbf{e}, \mathbf{e}_1, \mathbf{e}_2, \mathbf{e}_3, \cdots, \mathbf{e}_k$ are regular expressions, then replace it with the diagram

$$\longrightarrow \boxed{s_{k-1}} \begin{matrix} \xrightarrow{ee_1} \boxed{s_{i+1}} \\ \xrightarrow{ee_2} \boxed{s_{i+2}} \\ \xrightarrow{ee_3} \boxed{s_{i+3}} \\ \vdots \\ \xrightarrow{ee_w} \boxed{s_{i+n}} \end{matrix}$$

Example 3.20   Assume we begin with automaton

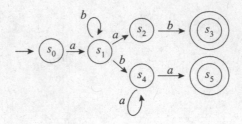

We then add a new terminal state T to get the automaton

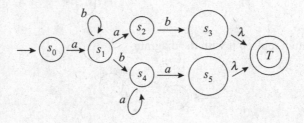

We now apply rule (2) to get the automaton

$$\rightarrow \boxed{s_0} \xrightarrow{a} \boxed{s_1} \begin{array}{c} \xrightarrow{ab} \\ \xrightarrow{ba^*a} \end{array} T$$

Apply rules (2) and (3) to get the automaton

$$\rightarrow \boxed{s_0} \begin{array}{c} \xrightarrow{ab^*ab} \\ \xrightarrow{ab^*ba^*a} \end{array} T$$

Hence the regular expression is $ab^*ab \vee ab^*ba^*a$.

Example 3.21   Given the automaton

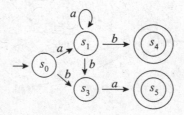

we go through the following steps

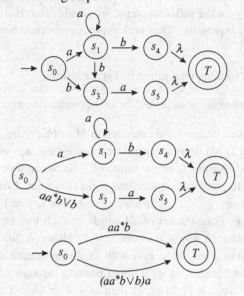

to get the regular expression $((aa^*b \vee b)a) \vee aa^*b$.

Example 3.22   Given the automaton

we go through the following steps,

to get the regular expression $(a \vee b)(a \vee b)^*(a \vee b)$. Note that the process is not unique and that by taking different steps, we would have had a different, but equivalent, regular expression. Thus both expressions would have described the same set.

We now give a formal proof of the following lemma

Lemma 3.4 *The language accepted by an automaton is regular.*

Proof Given a finite deterministic automaton $M = (\Sigma, Q, s_0, \Upsilon, F)$ we wish to show that $L = L(M)$, the language accepted by the automata, is regular. To do this we will express L as the union of a finite number of regular languages, and since the union of regular languages is regular, L is regular. Let Q contain n elements $q_1, q_2, \ldots, q_n$, where $s_0 = q_1$. For $i, j = 1$ to n and $k = 1$ to $n + 1$, let $R(i, k, j)$ be the set of all words w such that $(q_i, w) \vdash^* (q_j, \lambda)$ without passing through any q_m where $m \geq k$. However, q_i and q_j do not have this restriction. Thus there is a path in the automata such that M in state q_i reads w and is then in q_j without passing through m where $m \geq k$. Thus if $(q_i, w) \vdash^* (q_m, w') \vdash^* (q_j, \lambda)$ then $m < k$ or $m = i$ and $w' = w$ or $m = j$ and $w' = \lambda$. Hence the restriction is only on interior states of the path. Since there are only n states, $R(i, n + 1, j) = \{w : (q_i, w) \vdash^* (q_j, \lambda)$. Hence $L = \cup\{R(i, m, j) : j \in Q$.   □

To complete the proof, we need to show that $R(i, p, j)$ is regular for $1 \leq p \leq n + 1$. We do this using induction. If $p = 1$, then there are no interior states in the path so $R(i, p, j) = \{a \in \Sigma : \delta(q_i, a) = q_j\}$ if $i \neq j$ and $\{\lambda\} \cup \{a \in \Sigma : \delta(q_i, a) = q_j\}$ if $i = j$. Hence we have a finite set of elements of Σ and possibly λ in the set so it is a regular set.

Assume $R(i, k, j)$ is regular. The set of words $R(i, k + 1, j)$ can be defined as

$$R(i, k + 1, j) = R(i, k, j) \cup R(i, k, k)R(k, k, k)^* R(k, k, j)$$

where the path from q_i to q_j may not pass through a state q_m where $m \geq k$ or that passes along a path from q_i to q_k, then passes through zero or more paths from q_k to q_k and finally passes along a path from q_k to q_j. None of these paths passes along an interior state q_m where $m \geq k$. Since $R(i, k + 1, j)$ is formed using union, concatenation, and Kleene star of regular states, it is regular and hence L is regular.

Since we have now shown that every regular expression is accepted by an automaton and that the language accepted by an automaton is regular, we have proven Kleene's Theorem.

As a result of Kleene's Theorem, we discover two new properties about the regular languages:

Theorem 3.4 *If L_1 and L_2 are regular languages, then*

$$\Sigma^* - L_1 = \{x : x \in \Sigma^* \text{ and } x \notin L_1\}$$

and $L_1 \cap L_2$ are regular languages.

Proof To show $\Sigma^* - L_1$ is regular, let M_1 be a deterministic automaton for L_1. To construct the automaton for $\Sigma^* - L_1$, simply change all of the terminal states in M_1 to nonterminal states and all of the nonterminal states to terminal states. As a result, all words that were accepted because the automaton stopped in a terminal state, are no longer accepted and all words which were not accepted are now accepted since the automaton will now stop in a terminal state after reading this word.

To show that $L_1 \cap L_2$ is a regular language we simply use the set theory property that

$$L_1 \cap L_2 = \Sigma^* - ((\Sigma^* - L_1) \cup (\Sigma^* - L_2)).$$

This is most easily seen by thinking of Σ^* as the universe so that $\Sigma^* - L_1 = L_1'$ and the statement is simply $L_1 \cap L_2 = (L_1' \cup L_2')'$ which follows immediately from De Morgan's law and the fact that $L = L''$. Since the set of regular languages is closed under union and complement (first part of theorem), it is closed under intersection. $\qquad\square$

Exercises

(1) Let L_1 be the language accepted by the automaton

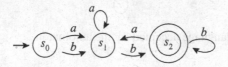

and L_2 be the language accepted by the automaton

→ s_0' $\underset{b}{\overset{a}{\rightleftarrows}}$ s_1' $\underset{b}{\overset{a}{\rightleftarrows}}$ s_2' $\circlearrowright$ a,b

with s_0' $\circlearrowleft$ b

(a) Construct the automaton which accepts the language $L_1 \cup L_2$.

(b) Construct the automaton which accepts the language $L_1 L_2$.

(c) Construct the automaton which accepts the languages L_1^* and the automaton which accepts L_2^*.

(2) Let L_1 be the language accepted by the automaton

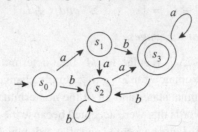

and L_2 be the language accepted by the automaton

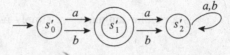

(a) Construct the automaton which accepts the language $L_1 \cup L_2$.

(b) Construct the automaton which accepts the language $L_1 L_2$.

(c) Construct the automaton which accepts the languages L_1^* and the automaton which accepts L_2^*.

(3) Let L_1 be the language accepted by the automaton

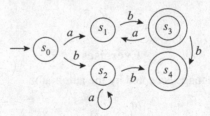

and L_2 be the language accepted by the automaton

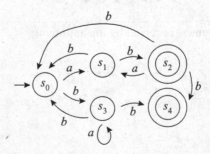

(a) Construct the automaton which accepts the language $L_1 \cup L_2$.

(b) Construct the automaton which accepts the language $L_1 L_2$.

(c) Construct the automaton which accepts the languages L_1^* and the automaton which accepts L_2^*.

(4) Let L_1 be the language accepted by the automaton

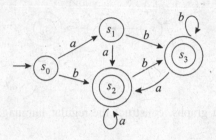

and L_2 be the language accepted by the automaton

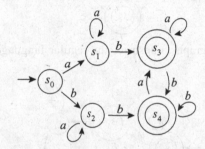

(a) Construct the automaton which accepts the language $L_1 \cup L_2$.

(b) Construct the automaton which accepts the language $L_1 L_2$.

(c) Construct the automaton which accepts the languages L_1^* and the automaton which accepts L_2^*.

(5) Using transition graphs, construct the regular language accepted by the automaton.

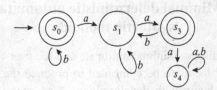

(6) Using transition graphs, construct the regular language accepted by the automaton.

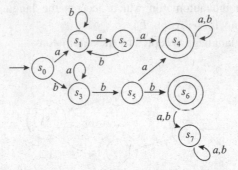

(7) Using transition graphs, construct the regular language accepted by the automaton

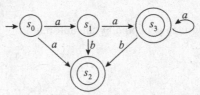

(8) Using transition graphs, construct the regular language accepted by the automaton

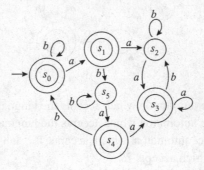

3.3   Minimal deterministic automata and syntactic monoids

In this section, we discuss minimal automata and the transformation monoid. We then show how they can be combined to produce the syntactic monoid of a language. We begin with the definition of an accessible deterministic automaton:

Definition 3.3 *The state s of an automaton M is **accessible** if there exists a word w in Σ^* so that if M reads the word w, it is then in state s. Equivalently the state s of an automaton M is **accessible** if there exists a word w in Σ^* so that in reading w, M passes through state s. An automaton M is **accessible** if every state in the automaton M is accessible.*

Intuitively a state s is accessible if one can begin at s_0 and follow a path of arrows to reach the state s.

For the moment, we shall adopt the following definition of a minimal automaton with a finite number of states.

Definition 3.4 *A deterministic automaton M is a **minimal** if the number of states in M is less than or equal to the number of states in any other deterministic automaton accepting the same language as M.*

Assume Υ is not empty. Obviously, if a state s in an automaton M is not accessible, it can be removed without changing the language accepted by M, but is M still deterministic? The answer is yes, if all of the states which are not accessible are removed. The reason is that if there is an a-arrow from s to s' and s' is not accessible, then s is not accessible, since any path from s_0 to s could be extended to a path from s_0 to s'.

An alternative way to deterine which states are accessible is to begin at the initial state s_0 and list all states to which there is an arrow from s_0. Call this list X. Enlarge X by adding any state to which there is an arrow from some state already in X. Iterate until X is no longer enlarged. The list X is then the set of accessible states.

A state h is co-accessible to a state g if there is no word of arrows from h to g. To find the co-accessible states, reverse the arrows and begin with the acceptance states.

A minimal state has no states that are not accessible or co-accessible. So they may be removed.

Therefore the first step in constructing a minimal deterministic automaton is to remove all states which are not accessible or co-accessible. Hence a minimal automaton is accessible and co-accessible.

We will now give an algorithm for constructing the minimal automaton which accepts a given language. The first begins with an automaton for the language and constructs the minimal automaton. The second begins with an automaton accepting the language and uses it to construct the minimal automaton.

At this point, we have several problems. The first is that removing states that are not accessible or not co-assessible does not necessarily give us a minimal automaton so we need to find out how to find a minimal automaton. (It does however tell us if the language is empty or consists only of the empty word.) The second is, for a given language, whether the minimal automaton is unique, even up to isomorphism.

At this point we will develop a method for developing a minimal automaton which has two advantages. It is developed using the language, so if we use this proceedure, and define a minimal automaton to be the one developed by this procedure, we will be able to consider **the minimal automaton**. The second advantage is that the procedure works for all languages although we already know that the automaton will have a finite number of states if and only if the language is regular.

In addition we shall develop a monoid called the syntactic monoid which will be discussed later (see Theorem 3.7). We introduce it now because the development of the minimal automaton and syntactic monoid are interrelated.

We now develop and define the **intrinsic automaton** and the **syntactic monoid** of an arbitrary language.

Definition 3.5 *As usual: Σ is the alphabet; Σ^* is the set of all strings over Σ and L is a language over Σ, i.e., a subset of Σ^*. Relative to the language L we define the **intrinsic** (or minimal) **automaton** M of L: The "states" of M are the equivalence classes defined, for each $x \in \Sigma^*$, by $[x] = \{y \in \Sigma^* \mid R(y) = R(x)\}$, where $R(x)$ is the set of "right contexts" accepted by x relative to L. Specifically: $R(x) = \{v \in \Sigma^* \mid xv \in L\}$. Each symbol a in Σ "acts on" the state $[x]$ by $[x]a = [xa]$ where $xa = \Upsilon(x, a)$. M has a specified Start state, 1, and a specified set of Acceptance states, $\{[x] \mid x \in L\}$. We may view M as a directed arrow-labeled graph having the states of L as its vertices, having directed edges $([x], a, [xa])$ where the second term is in Σ and is called the **label** of the arrow. The automaton M is considered to "recognize" each string in Σ^*, which produces a path from a Start state to an Acceptance state.*

The constructed M recognizes precisely those strings that are in L. A language L is regular if its automaton has only finitely many states. Since for each word in the language, there is a unique path from the start state to an acceptance state, the intrinsic automaton is minimal with regard to the above definition.

Definition 3.6 *The **syntactic monoid** S of L has as its elements the equivalence classes defined, for each $x \in \Sigma^*$, by $[[x]] = \{y \in \Sigma^* \mid LR(y) = LR(x)\}$, where $LR(x)$ is the set of "two-sided contexts" accepted by x relative to the language L. Specifically, $LR(x) = \{(u, v) \in \Sigma^* \times \Sigma^* \mid uxv \in L\}$ and S has an associative binary operation that is "well defined" by setting $[[x]][[y]] = [[xy]]$.*

Since [[1]] serves as a two-sided identity for this operation, S has the structure of a monoid, i.e., semigroup with an identity element. The partition of Σ^* into the classes [[x]] refines the partition of Σ^* into the classes [x] since $LR(x) = LR(y)$ implies $R(x) = R(y)$. Consequently when S is finite L is regular.

The action of Σ on the states of L extends, inductively, to an action of Σ^* on the states of L. Consequently each string $y \in \Sigma^*$ determines a function from the state set of L into itself defined by $[u]x = [ux]$. Two strings x and y determine the same function precisely if, for every $u \in \Sigma^*$, $[ux] = [uy]$. But this holds precisely if, for all $v \in \Sigma^*$, $uxv \in L$ if and only if $uyv \in L$. Thus x and y determine the same function precisely if $[[x]] = [[y]]$. When L is regular there can be only a finite number of functions from the state set of L into itself. Consequently when L is regular S is finite.

Summary For *every* language L we have an intrinsically associated automaton that recognizes the language and we have an associated syntactic monoid. The following are equivalent: (1) L is a regular language; (2) the intrinsic automaton for L has only finitely many states; and (3) the syntactic monoid of L is finite.

The word "intrinsic" is used because each language provides a unique automaton using this process (not just an isomorphism type – but one unique set of states, and arrows (or transitions). Thus if it is used there is no concern about isomorphism – the intrinsic automaton is 100% unique.

Now the question of isomorphism can come up (as it certainly will in elementary automata theory) when one uses some arbitrary automaton that recognizes the language and a different process for finding the minimal automaton. We shall develop another process and show that when using this process on any automaton accepting the language, the minimal automaton is isomorphic to the intrinsic automaton and hence the minimal automata developed for different automata for the same language produces isomorphic minimal automata.

We shall now use an algorithm for "collapsing pairs of states" (without altering the language being recognized) until no further collapsing is possible. Thus producing a minimal automaton.

Here is the procedure:

Procedure

Step 1 For each set of pairs of states $\{p, q\}$, determine if there is a string with length 0 that will take exactly one of these states into a final state (of course the other into a nonfinal state). In case of (the) length zero string, this just means, determine whether one of these states is final and the other is not. If so p and q can NEVER be collapsed without altering the language accepted. Mark this pair for "non-collapse"!

Step 2   For each remaining UN-MARKED pair $\{p, q\}$ and each symbol b in the alphabet, note $\{\Upsilon(p, b), \Upsilon(q, b))$. If $\Upsilon(p, b)$ and $\Upsilon(q, b)$ are distinct and the pair they form was Marked in the PREVIOUS round then p and q can NEVER be collapsed without altering the language recognized. Mark such pair for "non-collapse"!

Repeat step 2 until, when the step is completed no new pairs have been Marked.

Note that for each pair $\{p, q\}$ remaining unmarked at this stage: For any string s of symbols of the alphabet, $\Upsilon(p, s)$ and $\Upsilon(q, s)$ (starting in states p and q, the string s is read) must be either both final states or both non-final states.

Note that the following defines an equivalence relation in the set S of states of the original automaton: $p \sim q$ if $\{p, q\}$ is an unmarked pair.

Collapse the state set S of the original automaton onto the set $S/\sim$. A state in $S/\sim$ is final if it consists of final states of S. Each $\Upsilon(p, s) = q$ of the original automaton provides $\Upsilon'([p], s) = [q]$ in the (minimized) automata where $[p]$, $[q]$ are the $\sim$ equivalence classes containing p and q.

Example 3.23   Let M be the deterministic automaton

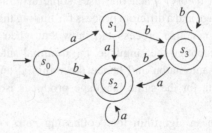

The unmarked pairs in step 1 are $\{s_0, s_0\}$, $\{s_1, s_1\}$, $\{s_2, s_2\}$, $\{s_3, s_3\}$, $\{s_0, s_1\}$, and $\{s_2, s_3\}$. The unmarked pairs in the first use of step 2 are $\{s_0, s_0\}$, $\{s_1, s_1\}$, $\{s_2, s_2\}$, $\{s_3, s_3\}$, and, $\{s_2, s_3\}$, since there is an a-arrow from s_0 to s_1 and an a-arrow from s_1 to s_2 and the states s_1 and s_2 are not in the unmarked pairs for step 0. Further uses of step 2 produce no new results.

The equivalences classes are $\{\{s_0\}, \{s_1\}, \{s_2, s_3\}\}$, and we are finished. In the graph shown below, only one element is picked from each equivalence class.

Therefore a minimal deterministic automaton is the automaton

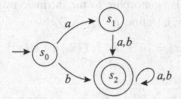

Example 3.24 Let M be the deterministic automaton

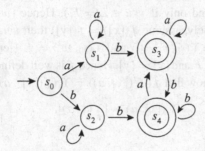

The unmarked pairs in step 1 are

$\{s_0, s_0\}, \{s_1, s_1\}, \{s_2, s_2\}, \{s_3, s_3\}, \{s_4, s_4\}, \{s_0, s_2\}, \{s_0, s_1\}$ and $\{s_1, s_2\}$.

The unmarked pairs in the first use of step 2 are

$\{s_0, s_0\}, \{s_1, s_1\}, \{s_2, s_2\}, \{s_3, s_3\}, \{s_4, s_4\}$ and $\{s_1, s_2\}$.

The second use of step 2 produces no new results so the equivalence classes are $\{\{s_0\}, \{s_1, s_2\}, \{s_3, s_4\}\}$, and we are finished.

Therefore a minimal deterministic automaton is the automaton where an element is picked from each equivalence class.

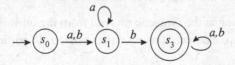

Now, one can see that this minimized version of the arbitrary automaton recognizing the language L is virtually identical with the intrinsic automaton of the language.

Theorem 3.5 *For a given regular language L, the two minimal reduced automaton developed above accepting language L are isomorphic.*

Proof $M = (\Sigma, Q, s_0, \Upsilon', F)$, the minimal reduced automaton developed by the collapsing method is isomorphic to the intrinsic minimal automaton. So $M_i = (\Sigma, Q_i, [1], \Upsilon_i, F_i)$. Define $f : Q \to Q_i$ by

$$f([x]) = \{w \in \Sigma^* : \Upsilon(x_0, w) \in [x]\}.$$

Thus

$$f([x]) = \{w \in \Sigma^* : \Upsilon(s_0, w) = x \text{ for } x \in [x]\}.$$

Assume $[x] = [y]$, then $\Upsilon(x, u) \in F$ if and only if $\Upsilon(y, u) \in F$ for $u, v \in \Sigma^*$. Let $f([x]) = [w]$ and $f([y]) = ([w'])$.

Then $wu \in L$ if and only if $w'u \in L(= F_i)$. Hence $[w] = [w']$ and f is well defined. Conversely, assume $f([x]) = f([y])$ then $wu \in L$ if and only if $w'u \in L(= F_i)$ where $\Upsilon(s_0, w) = x$ and $\Upsilon(s_0, w') = y$. Hence $\Upsilon(x, u) \in F$ if and only if $\Upsilon(y, u) \in F$ and $[x] = [y]$. Hence f is well defined and one-to-one.

Finally we must show that $f(\Upsilon'([x], a)) = \Upsilon_i(f([x]), a)$,

$$\Upsilon'([x], a) = [\Upsilon(x, a)],$$

and

$$\Upsilon_i(f([x]), a) = f([x])a.$$

Let $w \in f([x])$, then $\Upsilon(s_0, w) = x$ for $x \in [x]$. Let

$$\Upsilon(x, a) = y \in [\Upsilon(x, a)] = \Upsilon'([x]], a)$$

and $[y] = \Upsilon'([x]], a)$. Now $\Upsilon(s_0, wa) = y$, so

$$f([y])_i = [wa]_i = [w]_i a = f([x])a = [\Upsilon_i(f([x]), a)]$$

and so $f(\Upsilon'([x], a)) = \Upsilon_i(f([x]), a)$. $\qquad\square$

Corollary 3.1 *For a given regular language, all reduced automata which accept that language are unique up to isomorphism.*

Instead of looking at the syntactic monoid from the intrinsic point of view, as defined above we examine it using an automaton. In particular we look at minimal automata.

The **transformation monoid** of a deterministic automaton

$$M = (\Sigma, Q, s_0, \Upsilon, F)$$

is the image of a homomorphism φ from Σ^* to a submonoid T_M of the monoid of all functions from Q to Q. If $a \in \Sigma$, then $\varphi(a) = \bar{a}$ where for each $s_i \in Q$, $\bar{a}(s_i) = s_j$ if there is an a-arrow from s_i to s_j, i.e. $\Upsilon(s_i, a) = s_j$. If $a, b \in \Sigma$, then $\overline{ab} = \bar{a}\bar{b}$ where $\overline{ab}(s) = \bar{a}(\bar{b}(s))$. More specifically, for $u \in \Sigma^*$, $\bar{u}(s_i) = s_j$

if $(s_i, u) \vdash^* (s_j, \lambda)$. In other words, if the machine is in state s_i and reads u, then it is in state s_j.

Let M be the automaton

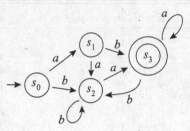

then

$$\bar{a}(s_0) = s_1 \quad \bar{a}(s_1) = s_2 \quad \bar{a}(s_2) = s_3$$
$$\bar{a}(s_3) = s_3 \quad \bar{b}(s_0) = s_2 \quad \bar{b}(s_1) = s_3$$
$$\bar{b}(s_2) = s_2 \quad \bar{b}(s_3) = s_2.$$

For convenience, permutation notation is used here although the functions are not usually permutations, since they are not one-to-one. Thus we have

$$\bar{a} = \begin{pmatrix} s_0 & s_1 & s_2 & s_3 \\ s_1 & s_2 & s_3 & s_3 \end{pmatrix} \text{ and } \bar{b} = \begin{pmatrix} s_0 & s_1 & s_2 & s_3 \\ s_2 & s_3 & s_2 & s_2 \end{pmatrix}$$

which we shall shorten to

$$\bar{a} = \begin{pmatrix} 0 & 1 & 2 & 3 \\ 1 & 2 & 3 & 3 \end{pmatrix} \text{ and } \bar{b} = \begin{pmatrix} 0 & 1 & 2 & 3 \\ 2 & 3 & 2 & 2 \end{pmatrix}.$$

By definition let $\bar{\lambda} = \begin{pmatrix} 0 & 1 & 2 & 3 \\ 0 & 1 & 2 & 3 \end{pmatrix}$. We now perform the following products:

$$\bar{a}\bar{b} = \begin{pmatrix} 0 & 1 & 2 & 3 \\ 1 & 2 & 3 & 3 \end{pmatrix} \begin{pmatrix} 0 & 1 & 2 & 3 \\ 2 & 3 & 2 & 2 \end{pmatrix} = \begin{pmatrix} 0 & 1 & 2 & 3 \\ 3 & 3 & 3 & 3 \end{pmatrix}$$

$$\bar{b}\bar{a} = \begin{pmatrix} 0 & 1 & 2 & 3 \\ 2 & 3 & 2 & 2 \end{pmatrix} \begin{pmatrix} 0 & 1 & 2 & 3 \\ 1 & 2 & 3 & 3 \end{pmatrix} = \begin{pmatrix} 0 & 1 & 2 & 3 \\ 3 & 2 & 2 & 2 \end{pmatrix}$$

$$\bar{a}\bar{a} = \begin{pmatrix} 0 & 1 & 2 & 3 \\ 1 & 2 & 3 & 3 \end{pmatrix} \begin{pmatrix} 0 & 1 & 2 & 3 \\ 1 & 2 & 3 & 3 \end{pmatrix} = \begin{pmatrix} 0 & 1 & 2 & 3 \\ 2 & 3 & 3 & 3 \end{pmatrix}$$

$$\bar{b}\bar{b} = \begin{pmatrix} 0 & 1 & 2 & 3 \\ 2 & 3 & 2 & 2 \end{pmatrix} \begin{pmatrix} 0 & 1 & 2 & 3 \\ 2 & 3 & 2 & 2 \end{pmatrix} = \begin{pmatrix} 0 & 1 & 2 & 3 \\ 2 & 2 & 2 & 2 \end{pmatrix}$$

$$\bar{a}\bar{b}\bar{b} = \begin{pmatrix} 0 & 1 & 2 & 3 \\ 2 & 3 & 3 & 3 \end{pmatrix} \begin{pmatrix} 0 & 1 & 2 & 3 \\ 2 & 2 & 2 & 2 \end{pmatrix} = \begin{pmatrix} 0 & 1 & 2 & 3 \\ 3 & 3 & 3 & 3 \end{pmatrix} = \bar{a}\bar{b}.$$

Continuing this process and letting $\gamma = \bar{a}\bar{b}$, $\delta = \bar{a}\bar{a}$, $\varepsilon = \bar{b}\bar{b}$, and $\zeta = \bar{b}\bar{a}$, the table for the transformation monoid T_M is seen to be

	$\bar{\lambda}$	$\bar{a}$	$\bar{b}$	γ	δ	ε	ζ
$\bar{\lambda}$	$\bar{\lambda}$	$\bar{a}$	$\bar{b}$	γ	δ	ε	ζ
$\bar{a}$	$\bar{a}$	δ	γ	γ	γ	γ	γ
$\bar{b}$	$\bar{b}$	ζ	ε	ε	ε	ε	ε
γ	γ	γ	γ	ε	γ	γ	γ
δ	δ	γ	γ	γ	γ	γ	γ
ε	ε	ε	ε	ε	ε	ε	ε
ζ	ζ	ε	ε	ε	ε	γ	γ

Example 3.25 Let M be the automaton

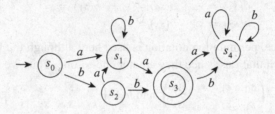

then

$$\bar{a} = \begin{pmatrix} 0 & 1 & 2 & 3 & 4 \\ 1 & 3 & 1 & 4 & 4 \end{pmatrix} \text{ and } \bar{b} = \begin{pmatrix} 0 & 1 & 2 & 3 & 4 \\ 2 & 1 & 3 & 4 & 4 \end{pmatrix}.$$

By definition let $\bar{\lambda} = \begin{pmatrix} 0 & 1 & 2 & 3 & 4 \\ 0 & 1 & 2 & 3 & 4 \end{pmatrix}$. Let

$$\gamma = \bar{a}\bar{b} = \begin{pmatrix} 0 & 1 & 2 & 3 & 4 \\ 1 & 3 & 4 & 4 & 4 \end{pmatrix} \qquad \delta = \bar{a}\bar{a} = \begin{pmatrix} 0 & 1 & 2 & 3 & 4 \\ 3 & 4 & 3 & 4 & 4 \end{pmatrix}$$

$$\varepsilon = \bar{b}\bar{b} = \begin{pmatrix} 0 & 1 & 2 & 3 & 4 \\ 3 & 1 & 4 & 4 & 4 \end{pmatrix} \qquad \zeta = \bar{b}\bar{a} = \begin{pmatrix} 0 & 1 & 2 & 3 & 4 \\ 1 & 4 & 1 & 4 & 4 \end{pmatrix}$$

$$\eta = \bar{a}\bar{a}\bar{b} = \begin{pmatrix} 0 & 1 & 2 & 3 & 4 \\ 3 & 4 & 4 & 4 & 4 \end{pmatrix} \qquad \theta = \bar{b}\bar{a}\bar{b} = \begin{pmatrix} 0 & 1 & 2 & 3 & 4 \\ 1 & 4 & 4 & 4 & 4 \end{pmatrix}$$

$$\vartheta = \bar{a}\bar{b}\bar{a} = \begin{pmatrix} 0 & 1 & 2 & 3 & 4 \\ 3 & 4 & 3 & 4 & 4 \end{pmatrix} \qquad \iota = \bar{a}\bar{b}\bar{b} = \begin{pmatrix} 0 & 1 & 2 & 3 & 4 \\ 4 & 3 & 4 & 4 & 4 \end{pmatrix}$$

$$\kappa = \bar{b}\bar{b}\bar{b} = \begin{pmatrix} 0 & 1 & 2 & 3 & 4 \\ 4 & 1 & 4 & 4 & 4 \end{pmatrix} \qquad \mu = abab = \begin{pmatrix} 0 & 1 & 2 & 3 & 4 \\ 3 & 4 & 4 & 4 & 4 \end{pmatrix}.$$

The table for the transformation monoid T_M is seen to be

	λ	$\bar{a}$	$\bar{b}$	γ	δ	ε	ζ	η	θ	ϑ	ι	κ	μ
λ	λ	$\bar{a}$	$\bar{b}$	γ	δ	ε	ζ	η	θ	ϑ	ι	κ	μ
$\bar{a}$	$\bar{a}$	δ	γ	η	η	ι	ϑ	η	μ	η	η	ι	η
$\bar{b}$	$\bar{b}$	ζ	ε	θ	η	κ	ζ	η	θ	η	η	κ	η
γ	γ	μ	ι	μ	η	ι	δ	η	μ	η	η	ι	η
δ	δ	η	η	η	η	η	η	η	η	η	η	η	η
ε	ε	ζ	κ	θ	η	κ	ζ	η	θ	η	η	κ	η
ζ	ζ	θ	θ	η	η	η	κ	η	η	η	η	η	η
η	η	η	η	η	η	η	η	η	η	η	η	η	η
θ	θ	η	η	η	η	η	η	η	η	η	η	η	η
ϑ	ϑ	η	μ	η	η	η	η	η	η	η	η	η	η
ι	ι	ϑ	ι	μ	η	ι	ϑ	η	μ	η	η	ι	η
κ	κ	ζ	κ	κ	η	κ	ζ	η	θ	η	η	κ	η
μ	μ	η	η	η	η	η	η	η	η	η	η	η	η

Theorem 3.6 *Let $M(\Sigma, Q, s_0, \Upsilon, F)$ be a minimal deterministic automaton and T_M be the transformation monoid for M, then T_M is finite.*

Proof Each element of T_M is a function from Q to Q. If Q contains n elements, then there are n^n possible functions from Q to Q. Therefore the order of M is less than or equal to n^n. $\qquad\square$

Theorem 3.7 *The syntactic monoid of a regular language L is isomorphic to the transformation monoid of the minimal deterministic automaton M that accepts L.*

Proof Since, by the discussion following Definition 3.6, the syntactic monoid can be considered to be the transformation monoid of the intrinsic minimal deterministic automaton, and all minimal deterministic automata are isomorphic to the intrinstic minimal deterministic automaton, the transformation monoid is isomorphic to the syntactic monoid. $\qquad\square$

We now examine some of the properties of the syntactic monoid of a language. Unlike the transformation monoid, as mentioned above, the syntactic monoid of a language also exists for languages that are not regular.

Definition 3.7 *Let ϕ be a homomorphism from Σ^* to a monoid Ω. A set $L \subseteq \Sigma^*$ is **recognized** by Ω if $\phi^{-1}\phi(L) = L$.*

Theorem 3.8 *Let $L \subseteq \Sigma^*$. The following conditions are equivalent.*

(i) L is a regular language.
(ii) The syntactic monoid $Syn(L)$ is finite.
(iii) L is recognized by a finite monoid Ω.

Proof (i)⇒(ii) If L is a regular language, then its syntactic monoid is isomorphic to the transformational monoid of the minimal automaton generating L and hence is finite.

(ii)⇒(iii) Assume ϕ is a homomorphism from Σ^* to $Syn(L)$. If $w \in L$ and $\phi(w) = \phi(w')$, then $uwv \in L$ if and only if $uw'v \in L$ for all $u, v \in \Sigma^*$. In particular $\lambda w \lambda \in L$ if and only if $\lambda w' \lambda \in L$. So $w \in L$, if and only if $w' \in L$. Therefore $\phi^{-1}\phi(L) = L$. Since $Syn(L)$ is finite, L is recognized by a finite monoid.

(iii)⇒(i) Assume L is recognized by a finite monoid Ω and let $\phi : \Sigma^* \to \Omega$. To show L is a regular language, we construct an automaton $M(\Sigma, Q, s_0, \Upsilon, F)$ that accepts L. Let $Q = \Omega$. Define $\Upsilon : \Sigma \times \Omega \to \Omega$ by $\Upsilon(a, m) = m\phi(a)$, for all $m \in \Omega$ and $a \in \Sigma$. Let $s_0 = 1$, the identity element of Ω and $F = \phi(\Omega)$. Then $w \in L(M)$ if and only if $\Upsilon(w, 1) \in \phi(\Omega)$ if and only if $w \in \phi^{-1}(\phi(\Omega)) = L$.

$\square$

Exercises

(1) Find the minimal automaton which accepts the same language as the automaton

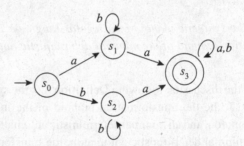

(2) Find the minimal automaton which accepts the same language as the automaton

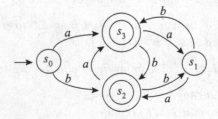

(3) Find the minimal automaton which accepts the same language as the automaton

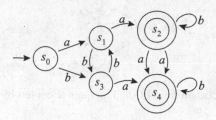

(4) Find the minimal automaton which accepts the same language as the automaton

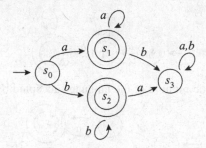

(5) Find the minimal automaton which accepts the language described by **aa*(b ∨ c)**.

(6) Find the minimal automaton which accepts the language described by **a(b ∨ c)*bb***.

(7) Find the minimal automaton which accepts the language described by **(abc)*(b ∨ c)**.

(8) Find the minimal automaton which acccpts the language described by **(a ∨ bc)c(ab)***.

(9) Find the syntactic monoid of the language accepted by the automaton

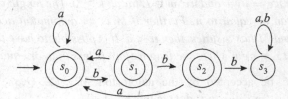

(10) Find the syntactic monoid of the language accepted by the automaton

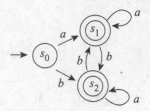

(11) Find the syntactic monoid of the language accepted by the automaton

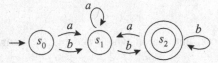

(12) Find the syntactic monoid of the language accepted by the automaton

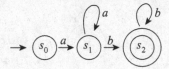

(13) Find the syntactic monoid of the language accepted by the automaton

3.4 Pumping Lemma for regular languages

We now show that certain languages are not regular languages. To do so we first prove a lemma known as the Pumping Lemma.

Lemma 3.5 (Pumping Lemma) *Let L be an infinite regular language. There exists a constant n such that if $z \in L$ and $|z| > n$, then there exists $u, v, w \in \Sigma^*$, $v \neq \lambda$ such that $z = uvw$ and $uv^k w \in L$ for all $k \geq 0$. The length of the string uw is less than or equal to n. Further if M is an automaton accepting the language L and M has q states, then $n < q$. It is possible to have the stronger statement that $z = uvw$ where the length of uv is less than or equal to q.*

Proof Let L be accepted by the automaton $M = (\Sigma, Q, s_0, \Upsilon, F)$. Let $\Upsilon(s_i, a_i) = s_{i+1}$ for $i = r$ to t; denote this by

$$(s_1, a_r a_2 a_3 \dots a_t) \vdash^* (s_{t+1}, \lambda).$$

Since L contains a word of length m, where $m > q$, say $w = a_1 a_2 a_3 \ldots a_m$. Note that if $(s_1, a_1 a_2 a_3 \ldots a_m) \vdash^* (s_m, \lambda)$, then s_m is an acceptance state. Since $m > q$, in reading w, M must pass through the same state twice. Therefore $(s_1, a_1 a_2 a_3 \ldots a_{j-1}) \vdash^* (s_k, \lambda)$ and $(s_1, a_1 a_2 a_3 \ldots a_{k-1}) \vdash^* (s_k, \lambda) =$ for some $j < k$ and both

$$(s_j, a_j a_{j+1} \ldots a_m) \vdash^* (s_m, \lambda) \quad \text{and} \quad (s_j, a_k a_{k+1} \ldots a_m) \vdash^* (s_m, \lambda).$$

Thus

$$(s_1, a_1 a_2 a_3 \ldots a_m) \vdash^* (s_m, \lambda) \text{ and } (s_1, a_1 a_2 \ldots a_{j-1} a_k a_{k+1} \ldots a_m) \vdash^* (s_m, \lambda).$$

Also $(s_j, a_j a_{j+2} \ldots a_{k-1}) \vdash^* s_j$, so in reading $a_j a_{j+2} \ldots a_{k-1}$, M returns to the same state and

$$(s_1, a_1 a_2 \ldots a_{j-1} (a_j a_{j+2} \ldots a_{k-1})^n a_k a_{k+1} \ldots a_m) \vdash^* (s_m, \lambda).$$

Letting $u = a_1 a_2 \ldots a_{j-1}$, $v = a_j a_{j+2} \ldots a_{k-1}$, and $w = a_k a_{k+1} \ldots a_m$, we have $uv^n w \in L$ for $n \geq 0$.

Since $|uw| < |uvw| = m$, if $|uw| > q$, we can repeat this process on uv until eventually we have $u'(v')^n w' \in L$ for $n \geq 0$ where $|u'w'| < q$. Let v be the first cycle in z produced by the same state being passed through twice when the automaton is reading z. Then the length of uv is less than or equal to q. Note that it is no longer true that the length of uw is less than q. $\square$

Using this lemma, we have the following theorem:

Theorem 3.9 *The language $L = \{a^n b^n : n \geq 1\}$ is not regular.*

Proof Assume $L = \{a^n b^n : n \geq 1\}$ is regular. Since L is infinite, there exist strings $u, v, w \in \Sigma^*$, $v \neq \lambda$ such that $uv^* w \subseteq L$. There are three possibilities. First $u = a^{m-k}$, $v = a^k$, and $w = b^m$ for some m. But then $a^{m-k} a^{2k} b^m = a^{m+k} b^m \in L$, which is a contradiction. Second, $u = a^m$, $v = b^k$, and $w = b^{m-k}$. By a similar argument, we reach a contradiction. Third $u = a^{m-k}$, $v = a^k b^r$, and $w = b^{m-r}$. But then $a^{m-k} a^k b^r a^k b^r b^{m-r} \in L$, which is a contradiction. Hence L is not regular. $\square$

Exercises

For each of the following sets, determine if the set is regular. If it is, describe the set with a regular expression. If it is not a regular set, use the Pumping lemma to show that it is not.

(1) $\{a^{2n} b^n : n \geq 1\}$.

(2) $\{a^n b^{2n} a^n : n \geq 1\}$.

(3) $\{(ab)^n : n \geq 1\}$.

(4) $\{a^n b^n a^n : n \geq 1\}$.

(5) $\{a^n b^m : m, n \geq 1\}$.

(6) $\{ww : w \in \Sigma^* \text{ and } |\Sigma| = 2\}$.

(7) $\{a^{2n} : n \geq 1\}$.

(8) $\{w \in \{a, b\}^* : w \text{ contains an equal number of } a\text{s and } b\text{s}\}$.

(9) $\{w \in \{a, b\}^* : w \text{ contains exactly four } b\text{s}\}$.

(10) $\{ww^R : w \in \{a, b\}^* \text{ and the length of } w \text{ is less that or equal to three}\}$.

(11) $\{ww^R : w \in \{a, b\}^*\}$.

(12) $\{wcw^R : w \in \{a, b, c\}^*\}$.

(13) $\{w\bar{w} : w \in (0, 1)^* \text{ and } \bar{w} \text{ is the 1s complement of } w\}$.

(14) $\{w \in \{a, b, c\}^* : \text{the length of } w = n^2 : n \geq 1\}$.

(15) $\{w \in \{a, b, c\}^* : \text{the length of } w \geq n \text{ for some } n \geq 1\}$.

(16) $\{w \in \{a, b\}^* : w \text{ contains more } a\text{s than } b\text{s}\}$.

3.5   Decidability

In this section we answer the questions

(1) Is there an algorithm for determining whether the language accepted by a finite automaton is empty?

(2) Is there an algorithm for determining whether two finite automata accept the same language?

(3) Is there an algorithm for determining whether two regular languages are the same?

(4) Is there an algorithm for determining whether a language accepted by an automaton is infinite?

The key to all of these questions is that they require the algorithm to be able to provide a yes or no answer. We are not concerned with the efficiency of the algorithm but only if within some finite length of time the algorithm can answer the question. Note that if an algorithm can determine that a statement is true (or false) within some bounded length of time, then the algorithm can determine whether the statement is true.

We begin with a proof of the first question although we can see that if we can answer the second question, then we can answer the first question. Given a language L, as an expression, we simply determine the automaton that accepts L and see if the language accepted is empty.

Theorem 3.10 *There is an algorithm for determining whether the language $M(L)$ accepted by a finite automaton is empty.*

Proof Let $M(L)$ have n states. Then $M(L)$ is empty if and only if s_0 is not an acceptance state and no string of length less than n is accepted since the shortest string accepted by $M(L)$ cannot enter a state twice. Since there are only a finite number of these strings, they can be checked. □

Theorem 3.11 *There is an algorithm for determining whether two finite automata accept the same language.*

Proof We already know that given automata M_1 and M_2 accepting languages $M_1(L)$ and $M_2(L)$, respectively, we can construct automata for accepting languages $M_1(L) \cap M_2(L)$, and $M_1(L) \cup M_2(L)$. Combining these constructions, we can find an automaton which accepts $(M_1(L) \cap M_2(L)') \cup (M_2(L) \cap M_1(L)')$, the symmetric difference of $M_1(L)$ and $M_2(L)$. But this set is empty if and only if $M_1(L) = M_2(L)$. Hence we use the previous theorem to determine whether $(M_1(L) \cap M_2(L)') \cup (M_2(L) \cap M_1(L)')$ is empty. □

Theorem 3.12 *There is an algorithm for determining whether two regular languages are the same.*

Proof Given expressions for L_1 and L_2, find the automata M_1 and M_2 so that $L_1 = M_1(L)$ and $L_2 = M_2(L)$. Now use the previous theorem to see if the two automata accept the same language. □

Before proving the next theorem, we need the following lemma.

Lemma 3.6 *Assume that an automaton M has n states. The language L accepted by M is infinite if and only if there is a word in L whose length is greater than n and less than $2n$.*

Proof First assume L is infinite. By the Pumping Lemma there exists $uv^m w \in L$ for all $m \geq 0$. Further if M is an automaton accepting the language L and M has n states, then $|uw|$, the length of the string uw, is less than or equal to n. Assume that after u is read, the machine is in state s. If while reading v, the machine returns to s, let v' be the string that is read when the machine first returns to s and $v'x = v$. Thus if we have

$$(s_0, uvw) \vdash^* (s, vw) \vdash^* (s, w) \vdash^* (s_2, \lambda),$$

replace it with

$$(s_0, uv'w) \vdash^* (s, v'w) \vdash^* (s, w) \vdash^* (s_2, \lambda).$$

Thus M reads the string $s_0, u(v')^n w$ for any nonnegative integer n. If while reading v', a state t is repeated, remove all of the states including one of the ts as well as the letters in v' that were read in this cycle. Thus we are simply removing all cycles in v'. Call this string v''. Since reading v'' uses no repeated states except s, the length of v'' is less than or equal to n. Thus the length of $uv''w$ is less than or equal to $2n$. If the length of $uv''w$ is less than or equal to n, there exists a least integer m so that the length of $u(v'')^n w$ is greater than n. Since the length of v'' is less than n, the length of $u(v'')^m w$ is less than $2n$.

Conversely in the proof of the Pumping Lemma, we showed that if there is a word in the language with length m greater than n, then for every positive integer r, the word $uv^r w \in L$, where v is nonempty. Hence L is infinite.

Theorem 3.13 *There is an algorithm for determining whether a language accepted by an automaton is infinite.*

Proof Let M have n states. Then $M(L)$ is infinite if and only if M accepts a string s with $n \le |s| \le 2n$. Since there are only a finite number of such words, check each of them to see if they are accepted by the automata.

Theorem 3.14 *There is an algorithm for determining whether a language is finite.*

Proof Using the proof of the previous theorem, if there is no string s accepted by M, with $n \le |s| \le 2n$, then $M(L)$ is finite (where we include the empty set and the set containing only the empty word as finite sets).   □

Theorem 3.15 *There is an algorithm determining whether a language $L_1 \subseteq L_2$.*

Proof We already know that there is an automaton that accepts $L_1 \cap L_2'$, which is empty if and only if $L_1 \subseteq L_2$.   □

Exercises

(1) Prove there is an algorithm for determining if regular language $M(L) = \Sigma^*$.
(2) Prove there is an algorithm for determining if a regular language $M(L)$ contains a word that contains a given letter of the alphabet.
(3) Prove there is an algorithm for determining if every letter in the alphabet is contained in some word in a regular language L.

(4) Prove that for a positive integer n, there is an algorithm for determining if a regular language contains a word with length less than n that contains a given letter of the alphabet.

(5) Prove that for a positive integer n, there is an algorithm for determining if every letter in the alphabet is contained in some word with length less than n in a regular language L.

(6) Prove there is an algorithm for determining if a regular language contains a word that begins with a given letter of the alphabet.

(7) Prove there is an algorithm for determining if there is a word in a regular language L of even length.

(8) Prove that for any integer k there is an algorithm for determining if there is a word in a regular language L of length mk for some m.

(9) Prove that for a regular language L, it is possible to determine if $\Sigma^* - L$ is finite.

3.6 Pushdown automata

In the previous section we mentioned that the set $\{a^n b^n : n$ is a positive integer$\}$ is not a regular language. Therefore it cannot be accepted by an automaton. Intuitively, the problem is that after the automaton has read the as in a word, it cannot remember how many it has read, so it does not know how many bs it should read. The automaton basically needs a memory so that it can remember the letters it has read. A **pushdown automaton** or **PDA** is essentially an automaton together with a very simple memory. The memory is called a **pushdown stack**. Associated with the stack is a set of symbols called the **stack symbols**. A stack symbol may be placed on the stack. This process is called **pushing** the symbol onto the stack. If x is a stack symbol, then **push** x simply means x is placed on the stack. The top symbol may also be removed from the stack. This is the last symbol placed on the stack. Since the last symbol placed in the stack is the first out, the stack is said to have the **LIFO** (last in–first out) property. Thus the symbols are removed from the stack in reverse order from the order they were put in the stack. The process of removing the top symbol from the stack is called **popping** the stack. If x is a stack symbol then **pop** x simply means that when the stack is popped, the symbol x is removed if it is on top of the stack. The purpose of the stack is to allow the PDA to remember the letters in the word that it has read so that it can duplicate them or replace them with other letters.

Assume that the word to be read is placed on a tape. The tape is divided into little squares with the letters of the word in the first squares. The rest of the tape is considered to be blank. Since the words may be arbitrarily long, it is best to use an infinite tape. These may have to be custom made. One of the advantages of mathematics is that mathematical structures do not usually have to be actually constructed.

The PDA, beginning at the left, reads a letter at a time in the same manner as a standard automaton. The PDA may read a letter from the tape or pop (remove from the top) and read a symbol from the stack or both. Depending on its current state and the symbol(s) read, the PDA may change state, push a symbol in the stack, or both.

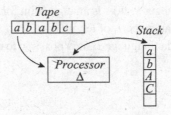

We now define a PDA more formally.

Let $\Sigma^\lambda = \Sigma \cup \{\lambda\}$ and $I^\lambda = I \cup \{\lambda\}$.

Definition 3.8 *A **pushdown automaton** is a sextuple*

$$M = (\Sigma, Q, s, I, \Upsilon, F)$$

where Σ is a finite alphabet, Q is a finite set of states, s is the initial or starting state, I is a finite of stack symbols, Υ is the transition relation and F is the set of acceptance states. The relation Υ is a subset of

$$((\Sigma^\lambda \times Q \times I^\lambda) \times (Q \times I^\lambda)).$$

Thus the relation reads a letter from Σ^λ, determines the state, and reads a letter from I^λ. It then changes state or remains in the same state and gives a letter of I^λ as output. Similar to the automata, the letter of a word is removed when it is read. The top letter on the stack is also removed when it is read. As discussed above, we say it is popped from the top of the stack. The letter of I produced by the relation is placed on top of the stack or pushed on the stack as discussed above. A word is **accepted** by the PDA if and only if after beginning in the start state, with an empty stack, the word is read, if possible, the machine is in an acceptance state, and the stack is empty. If all of the above do not occur, then the word is **rejected**. The language consisting of all words accepted by the pushdown automaton M is denoted by $M(L)$.

Elements of Υ have the following rules:

$((a, s, E), (t, D))$ In state s, a is read and E is popped, go to state t and push D.

$((a, s, \lambda), (t, D))$ In state s, a is read, go to state t and push D.

$((\lambda, s, \lambda), (s, D))$ In state s, push D.

$((a, s, E), (t, \lambda))$ In state s, and a is read, pop E and go to state t.

$((\lambda, s, E), (s, \lambda))$ In state s, pop E.

$((a, s, \lambda), (t, \lambda))$ In state s, read a and go to state t.

$((a, s, \lambda), (s, \lambda))$ In state s, read a.

$((\lambda, s, \lambda), (t, \lambda))$ Move from state s to state t.

Definition 3.9   *M is a **deterministic PDA** if $\Upsilon \subseteq ((\Sigma^\lambda \times Q \times I^\lambda) \times (Q \times I^\lambda))$ has the property that if $((s, a, c), (s', c'))$ and $((s, a, c), (s'', c'')) \in F$ then $s' = s''$ and $c' = c''$.*

Note that this definition differs between texts.

Since the requirement that M is a deterministic PDA restricts the languages that it accepts, we will not consider the deterministic PDA.

Although it seems a severe restriction, any language accepted by a PDA can be accepted by a PDA with only two states, which we will call s and t. The automaton leaves the first state before it reads the first letter and while the stack is still empty. The second state is then the terminal state. Often it is simpler or more convenient to use more states. An example of this will be shown in the examples.

As with the regular automaton, we will show the PDA graphically. The PDA will be shown as a flow chart using only the instructions start, read, push, pop, and accept. It will be obvious that the flow charts below describe PDAs. We shall not try to prove that every PDA has a flow chart. Each edge of a flow chart has a state associated with it. For example in the following figure, (t) on the edge indicates that at that point on the chart, we are in state t. We could put the state with each edge, but we only do so when the state is changed. Thus the state is determined by the location on the flow chart. When only two states are used, including the start command which takes the PDA to the state t, the state will not be indicated. The symbol indicates start and switch to state t. The symbol

indicates start, push S, and switch to state t. The symbol

indicates read a. The symbol

indicates pop a. The symbol

indicates push a. Finally, the symbol

indicates accept if the word has been read, the machine is in an acceptance state, and the stack is empty. Thus the diagram

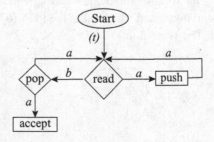

allows the PDA to read a and then push it or read b and pop a if it is on the stack. Thus every time a b is read, it removes an a which has been read and placed in the stack. In this example the alphabet and the stack symbols will both consist of a, and b. If, at any time, there were more bs read than as in the stack, there would be no a in the stack to remove and the PDA could not continue. If the number of as is equal to the number of bs when the word is read, then the stack will be empty. A word is accepted if, after popping a, the word has been read and the stack is empty. Therefore this PDA accepts words which have the same number of as and bs provided that, for every b in the word, the string preceding it contains more as than bs. For example consider the word $aababb$. We can trace its path with the following table:

instruction	stack	tape
start	λ	$aababb$
read	λ	$ababb$
push a	a	$ababb$
read	a	$babb$
push a	aa	$babb$
read	aa	abb
pop	a	abb
read	a	bb
push a	aa	bb
read	aa	b
pop	a	b
read	a	λ
pop	λ	λ
accept	λ	λ

Example 3.26   The PDA

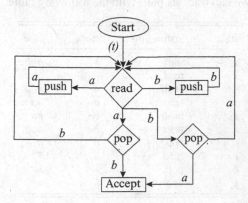

accepts words containing the same number of *a*s and *b*s. Consider the word *abba*. We can trace its path with the following table:

instruction	stack	tape
start	λ	*abba*
read	λ	*bba*
push *a*	*a*	*bba*
read	*a*	*ba*
pop	λ	*ba*
read	λ	*a*
push *b*	*b*	*a*
read	*b*	λ
pop	λ	λ
accept	λ	λ

In the following example, three states are used. A move to a new state is indicated in the diagram by an arrow for which there is no loop or are no return arrows.

Example 3.27   The PDA

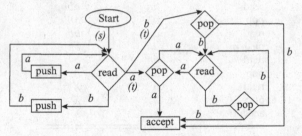

accepts words ww^R where w^R is the word w reversed. We read the first half of the word and then switch states to read the second half of the word. Consider the word *abba*. We can trace its path with the following table:

state	instruction	stack	tape
s	start	λ	*abba*
s	read	λ	*bba*
s	push *a*	*a*	*bba*
s	read	*a*	*ba*
s	push *b*	*ba*	*ba*
t	read	*ba*	*a*
t	pop	*a*	*a*
t	read	*a*	λ
t	pop	λ	λ
t	accept	λ	λ

Exercises

(1) Which of the following words are accepted by the following pushdown automaton M_1?

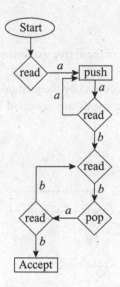

(a) *abbb*
(b) *aabbb*
(c) *aabbbbb*
(d) *aaabbb*
(e) *aabab*
(f) *aaabbbb*.

(2) Use a table to trace each of the above words through the pushdown automaton M_1.

(3) What is the language accepted by the pushdown automaton M_1?

(4) Which of the following words are accepted by the following pushdown automaton M_2?

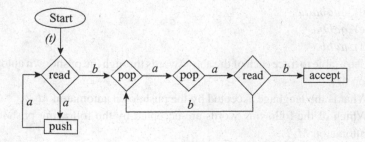

    (a) *abb*

    (b) *aabbaaa*

    (c) *aabbbaa*

    (d) *aaabaaa*

    (e) *aabba*

    (f) *aabb*.

(5) Use a table to trace each of the above words through the pushdown automaton M_2.

(6) What is the language accepted by the pushdown automaton M_2?

(7) Which of the following words are accepted by the following pushdown automaton M_3?

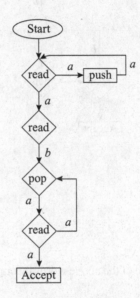

    (a) *abb*

    (b) *aabbaaa*

    (c) *aabbbaa*

    (d) *aaabaaa*

    (e) *aabba*

    (f) *aabb*.

(8) Use a table to trace each of the above words through the pushdown automaton M_3.

(9) What is the language accepted by the pushdown automaton M_3?

(10) Which of the following words are accepted by the following pushdown automaton M_4?

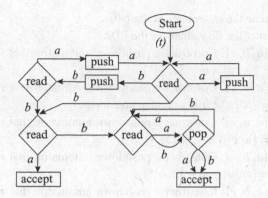

(a) *abb*

(b) *bb*

(c) *aabbbaaa*

(d) *abbbaa*

(e) *aabba*

(f) *aabb*.

(11) Use a table to trace each of the above words through the pushdown automaton M_4.

(12) What is the language accepted by the pushdown automaton M_4?

(13) Given a pushdown automaton $M = (\Sigma, Q, s_0, I, \Upsilon, F)$ where $\Sigma = I = \{a, b\}$, $Q = \{s_0, s_1, s_2\}$, $F = \{s_2\}$, and Υ has the following relations:

$((a, s_0, \lambda), (s_1, a))$	In state s_0, a is read, go to state s_1 and push a
$((b, s_0, \lambda), (s_1, b))$	
$((a, s_1, \lambda), (s_1, a))$	
$((b, s_1, \lambda), (s_1, b))$	
$((a, s_1, \lambda), (s_2, \lambda))$	
$((a, s_2, a), (s_2, \lambda))$	
$((b, s_2, b), (s_j, \lambda))$	

(a) Complete the statements in the table.

(b) Construct the flow chart for the PDA.

(14) Given a pushdown automaton $M = (\Sigma, S, s_0, I, \Upsilon, F)$ where $\Sigma = I = \{a, b\}$, $Q = \{s_0, s_1, s_2\}$, $F = \{s_2\}$, and Υ has the following relations:

$((a, s_0, \lambda), (s_1, a))$	In state s_0, a is read, go to state s_1 and push a
$((b, s_0, \lambda), (s_1, b))$	
$((a, s_1, a), (s_1, b))$	
$((a, s_1, b), (s_1, b))$	
$((a, s_1, a), (s_2, a))$	
$((b, s_1, a), (s_j, \lambda))$	
$((a, s_2, a), (s_2, a))$	
$((b, s_2, a), (s_j, \lambda))$	

(a) Complete the statements in the table.

(b) Construct the flow chart for the PDA.

(15) Let $\Sigma = \{a, b, c\}$. Construct a pushdown automaton that reads the language $L = \{wcw^r : w \in \{a, b\}^*\}$.

(16) Let $\Sigma = \{a, b, c\}$. Construct a pushdown automaton that reads the language $L = \{a^n cb^n : n \text{ is a nonnegative integer}\}$.

(17) Let $\Sigma = \{a, b, c\}$. Construct a pushdown automaton that reads the language $L = \{ww^r : w \in \{a, b\}^*\}$.

(18) Let $\Sigma = \{a, b, c\}$. Construct a pushdown automaton that reads the language $L = \{wcw^r : w \in \{a, b\}^*\}$.

(19) Let $\Sigma = \{a, b, c\}$. Construct a pushdown automaton that reads the language $L = \{w : \text{The number of } as \text{ in } w \text{ is equal to the sum of the number of } bs \text{ and } cs\}$.

(20) Let $\Sigma = \{a, b\}$. Construct a pushdown automaton that reads the language $L = \{w : \text{The number of } as \text{ in } w \text{ is equal to twice the number of } bs \text{ or the number of } bs \text{ in } w \text{ is equal to three times the number of } as\}$.

(21) Given two pushdown automata

$$\Gamma = (N, \Upsilon, S, P)$$

and

$$\Gamma' = (N', \Upsilon', S', P')$$

over the same alphabet Σ and accepting languages L and L' respectively,

(a) Describe how to construct a pushdown automaton Γ_1 that accepts the language $L \cup L'$.

(b) Construct a pushdown automaton Γ_1 that accepts the language $L \cup L'$ where L is the language accepted by the automaton in Example 3.26 and L' is the language accepted by the automaton in Example 3.27.

(22) Given two pushdown automata

$$\Gamma = (N, \Upsilon, S, P)$$

and

$$\Gamma' = (N', \Upsilon', S', P')$$

over the same alphabet Σ and accepting languages L and L' respectively,

(a) Describe how to construct a pushdown automaton Γ_2 that accepts the language LL'.

(b) Construct a pushdown automaton Γ_2 that accepts the language LL' where L is the language accepted by the automaton in Example 3.26 and L' is the language accepted by the automaton in Example 3.27.

(23) Given a pushdown automaton $\Gamma = (N, \Upsilon, S, P)$ over the alphabet Σ and accepting language L,

    (a) Describe how to construct a pushdown automaton Γ_3 which accepts the language L^*.

    (b) Construct a pushdown automaton Γ_3 that accepts the language $L \cup L'$ where L is the language accepted by the automaton in Example 3.26 and L' is the language accepted by the automaton in Example 3.27.

(24) Given two pushdown automata

$$\Gamma = (N, \Upsilon, S, P)$$

and

$$\Gamma' = (N', \Upsilon', S', P')$$

over the same alphabet Σ and accepting languages L and L' respectively,

    Construct a pushdown automaton Γ_4 that accepts the language $L \cup L'$ where L is the language accepted by the automaton in Example 3.26 and L' is the language accepted by the automaton in Example 3.27.

3.7   Mealy and Moore machines

Previously, we defined a deterministic automaton, a device which only accepts or recognizes words of a language of Σ^*. We now produce two machines which are similar to deterministic automata, but produce output.

The first machine we introduce is called a **Moore machine**, created by E. F. Moore[30] and is denoted by $(\Sigma, A, S, s_0, \Upsilon, \phi)$. It also has a finite set of states S including a starting state s_0. It contains two alphabets Σ and A. The first is the alphabet of input characters to be read by the machine. The second is the alphabet of output characters produced by the machine. The Moore machine retains the transition function $\Upsilon : S \times \Sigma \to S$ of the finite state automaton. It also contains an output function $\phi : S \to A$. In the operation of a Moore machine, the output is first produced using the output function ϕ before the transition function F is used to read the input and change states. Imitating the deterministic automaton, the Moore machine reads each element of a string w of characters of Σ until it has read the entire string. During this process, it produces output consisting of a string of characters of A. Since the Moore machine produces output $\phi(s_0)$ before the first input character is read and produces output from the last state reached before the transition function tries and fails to read input, the output string contains one more character than the input string. Also since $\phi(s_0)$ is always executed first, each output string must begin with

$\phi(s_0)$. As with the deterministic automata, we say a Moore machine **reads** a symbol a of the alphabet Σ to indicate that the letter a is used as input for the function Υ. Similarly, in state s_i, if the output is $\phi(s_i)$, we shall say that the machine **prints** the value $\phi(s_i)$, although the output may be used for an entirely different purpose. Thus one may envision a Moore machine reading a string in Σ from a tape and printing a string in $A*$ on the tape or on another tape.

As with the finite state automaton, we shall illustrate the Moore machine using a finite state diagram. As in the deterministic automaton, if $\Upsilon(s_i, a) = s_j$, this is represented by

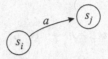

If $\phi(s_i) = z$, this is represented by

so that both s_i and $\phi(s_i)$ are represented inside the vertices of the diagram.

In the diagram

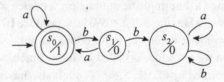

$\Sigma = \{a, b\}$, $A = \{0, 1\}$, $S = \{s_0, s_1, s_2\}$, Υ is given by the table

F	s_0	s_1	s_2
a	s_0	s_0	s_2
b	s_1	s_1	s_2

and ϕ is given by the table

s	$\phi(s)$
s_0	1
s_1	0
s_2	0

Given the input string *aba*, the machine first prints the value $\phi(s_0) = 1$. It then reads *a* and remains in state $\Upsilon(a, s_0) = s_0$. It then prints $\phi(s_0) = 1$. Next it reads *b* and travels to state $\Upsilon(b, s_0) = s_1$. It then prints $\phi(s_1) = 0$. Next it reads *a* and travels to state $\Upsilon(a, s_1) = s_0$. It then prints $\phi(s_0) = 1$. Since there is no more input, operations cease. The result is the output string 1101. The input string *aabab* produces the output string 111010. The input string *baab* produces the output string 10110.

Note that the Moore machine we have produced is actually the finite automaton

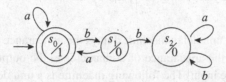

except that we have added ϕ with the property that $\phi(s_i) = 0$ if s_i is not an acceptance state and $\phi(s_i) = 1$ if s_i is an acceptance state. When we do this, the last character printed will be 1 if and only if the input is accepted by the finite automaton. Thus since the outputs for *aba* and *ababa* are 1101 and 110101 respectively, *aba* and *ababa* are accepted by the automaton. Using this procedure we can "duplicate" any finite automaton with a Moore machine where a word is accepted only if the last character output is 1. It may also be observed that whenever a 1 appears in the output, the initial string of input which has been read at that point is accepted by the finite automaton since the state at that point is an acceptance state. For example, in the above example input *aabaabbab* produces output 0001001001, so *aab*, *aabaab*, and *aabaabbab* are all accepted by the automaton. Since $\phi(s_0) = 1$ the empty word is also accepted. In general, the number of 1s in the output of a Moore machine which "duplicates" a finite automaton is the number of initial strings of the input which are accepted by the finite automaton.

Example 3.28 The automaton

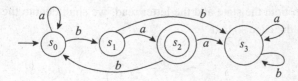

has corresponding Moore machine

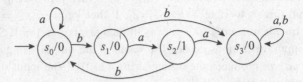

Input *babbab* produces output 0010010 so substrings *ba* and *babba* are accepted by the automaton. Since the input *ababbbau* produces output 000100000, the only substring accepted by the automaton is *aba* since only one 1 occurs.

Example 3.29   A unit delay machine delays the appearance of a bit in a string by one bit. Hence the appearance of a character in the output is preceded by one character in the input. The following machine is a unit delay machine.

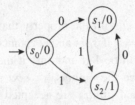

So far, we have primarily shown that a Moore machine may be used to "duplicate" a finite automaton. This is only one of the uses of a Moore machine. However, any task performed by a Moore machine can be performed by another machine called a Mealy machine and conversely. In most cases the task is more easily shown using a Mealy machine.

The **Mealy machine** also contains an output function, however, the input is an edge rather that a state. Since the edge depends on the state and the input, the output function δ "reads" a letter of $a \in \Sigma$ and the current state and prints out a character of the output alphabet. Hence δ is a function from $S \times \Sigma$ to A. More formally a Mealy machine is a sextuple $M_e = (\Sigma, A, S, s_0, \Upsilon, \delta)$ where Σ, A, S, s_0, and Υ are the same as in the Moore machine and $\delta : S \times A \to \Sigma$. The Mealy machine is also best illustrated using a finite state diagram. Since δ depends on both the state and the letter read, we shall denote the output by placing it on the edge so that

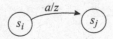

corresponds to $\Upsilon(s_i, a) = s_j$ and $\delta(s_i, a) = z$. Note that, unlike in the Moore machine, the output occurs after the input is read. Hence for every letter of input, there is a character of output.

Consider the Mealy machine

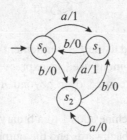

The functions Υ and δ are given by tables

Υ	s_0	s_1	s_2
a	s_1	s_2	s_2
b	s_2	s_0	s_1

and

δ	s_0	s_1	s_2
a	1	1	0
b	0	0	0

Given the input string $aaabb$, a is read, 1 is printed, and the machine moves to state s_1. The second a is read, 1 is printed, and the machine moves to state s_2. The third a is read, 0 is printed, and the machine remains at state s_2. The letter b is read, 0 is printed, and the machine moves to state s_1. Finally, b is read, 0 is printed, and the machine reaches state s_0. Thus input $aaabb$ produces output 11000.

Example 3.30 The Mealy machine

simply converts every a in the string to x, every b to y, and every c to z. Thus $aabbcca$ is converted to $xxyyzzx$.

Example 3.31 The **1s complement** of a binary string converts each 1 in the string to a 0 and each 0 to a 1. It is given by the state diagram

Example 3.32 If 1 is added to the 1s complement of a binary string of length n, we obtain the **2s complement** used to express the negative of an integer if we discard any number carried over beyond n digits. Thus $1111 + 1 = 0000$.

The following Mealy machine adds 1 to a binary string in this fashion. The input string must be read in backwards and the output is printed out backwards so the unit digit is read first. The stage diagram

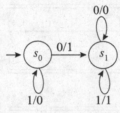

describes the Mealy machine. In this diagram, s_1 is the state reached if there is no 1 to carry when adding the digits. The state s_2 is reached if there is a 1 to carry when adding the digits. Let 1101 be the number in reverse. (Hence the actual number is 1011.) First input 1 is read. The output is 0 and the machine is in state s_2. (This corresponds to $1 + 1 = 10$ so 0 is output and 1 is carried.) Now input 1 is read. The output is 0 and the machine remains in state s_2. (This corresponds to $1 + 1 = 10$ so 0 is output and 1 is carried.) Next 0 is input. The output is 1 and the machine moves to state s_1. (This corresponds to $1 + 0 = 1$ so 1 is output and nothing is carried.) Finally 1 is input. The output is 1 and the machine remains in state s_1. (This corresponds to $1 + 0 = 1$ so 1 is output and nothing is carried.) Thus the output is 0011 and the number is 1100.

Example 3.33 The Mealy machine M_+ adds two signed integers. The signed integer m is subtracted from the signed integer n by adding n to the 2s complement of m. Thus M_+ can also be used for subtraction by first using the machine in the previous example to find the 2s complement of the number to be subtracted. Assume $a_n, a_{n-1}, \ldots, a_2, a_1$ and $b_n, b_{n-1}, \ldots, b_2, b_1$ are the two strings to be added. We again assume that the two strings to be added are read in reverse so the first two digits to be input are a_1 and b_1, followed by a_2 and $b_2, \ldots$, followed by a_n and b_n. We shall consider the pair of digits to be input as ordered pairs, so that (a_1, b_1) is the first element of input. The machine M_+ is

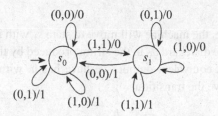

The machine is in state s_0 when no 1 has been carried in adding the previous input and is in state s_1 when a 1 has been carried in the addition. Assume that 0101 and 1101 are added. First $(1, 1)$ is read, so the machine moves to s_1 and prints 0. Next $(0, 0)$ is read, so the machine moves to s_0 and prints 1. Then $(1, 1)$ is read, so the machine returns to s_1 and prints 0. Finally $(1, 0)$ is read, so the machine remains at s_1 and prints 0. Note that the 1, if it exists, which is carried from adding the last two digits is discarded. Thus the sum of 0101 and 1101 is 0010.

Earlier in this section, we implied that Moore machines and Mealy machines were equivalent in the sense that every Moore machine could be duplicated by a Mealy machine and conversely. More specifically, given a Moore machine, there is a Mealy machine which will produce output equivalent to the Moore machine when given the same input. Conversely given a Mealy machine, there is a Moore machine which will produce the output equivalent to the Mealy machine when given the same input.

We first need to specify what we mean by equivalent output since a Mealy machine always has one less symbol of output than the Moore machine. A string of output of a Mealy machine is **equivalent** to a string of output of a Moore machine if it is equal to the substring of the Moore machine excluding the first symbol $\phi(s_0)$. Thus if the Moore machine produced output 010010101, the equivalent output from the Mealy machine would be 10010101.

The transformation from the Moore machine to an equivalent Mealy machine is the simplest. With the transition

in a Moore machine, given input c, the character a_0 will be printed, the machine will move to state s_1, and a_1 will next be printed. In the transition

of a Mealy machine, the machine will move to state s_1 with input c and a_1 will be printed. Since we disregard a_0 in the string produced by the Moore machine in our definition of equivalent output, we have begun with the same output. Assume that we have the transition

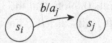

in a Moore machine and a_i has already been printed. Input b moves the machine to state s_j, and the next output will be a_j. The corresponding transition in the Mealy machine is

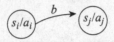

which produces the same transition and output.

Example 3.34 The Mealy machine corresponding to the Moore machine

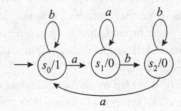

is

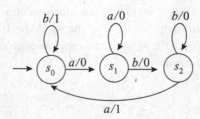

In transforming a Mealy machine to a Moore machine, we have to consider the problem where arrows into a given state produce different output. Consider the following example:

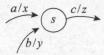

In a Moore machine, the state s produces unique output so it cannot produce both x and y as output. We solve this by making two copies of s

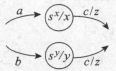

One will produce x as output and the other y as output as follows. Obviously both machines produce output x with input a and output y with input b. For simplicity, we shall simplify s^x/x to s/x and s^y/y to s/y noting that they are different states.

In general, for each state s, except the starting state, in a Mealy machine and for each output symbol z, we shall produce a copy s/z of the state s. This may result in some overkill since in the above example, if the output symbols were x, y, and z, we would not have needed state s/z since there was no arrow entering s with output z. We begin with initial state s_0 and give it an arbitrary output variable x_0 from the set of output variables since it is not used in producing output equivalent to the Mealy machine. If we have

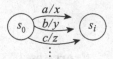

in the Mealy machine, we replace it with

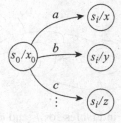

in the Moore machine. For other states, we replace

with

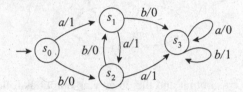

We produce the same output at each step for both machines.

Thus the machine equivalent to

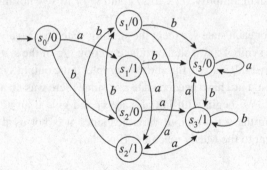

is strcj.eps

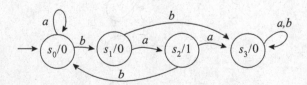

Exercises

(1) Let the Moore machine $M_o = (\Sigma, A, S, s_0, \Upsilon, \phi)$ be given by the diagram

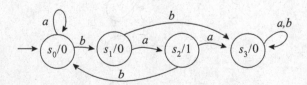

Describe A, Σ, and S. Find tables for F and ϕ.

(2) Let the Moore machine $M_o = (\Sigma, A, S, s_0, \Upsilon, \phi)$ be given by the diagram

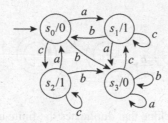

Describe A, Σ, and S. Find tables for Υ and ϕ.

(3) Let the Moore machine $M_o = (\Sigma, A, S, s_0, \Upsilon, \phi)$ be given by the diagram

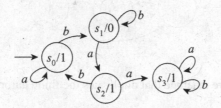

(a) Find the output with input *bbabab*.
(b) Find the output with input *aaabbaba*.
(c) Find the output with input *bbbaaa*.
(d) Find the output with input λ, the empty word.

(4) Let the Moore machine $M_o = (\Sigma, A, S, s_0, \Upsilon, \phi)$ be given by the diagram

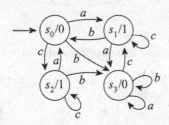

(a) Find the output with input *abcabca*.
(b) Find the output with input *bbbaaacc*.
(c) Find the output with input *aabbccaa*.
(d) Find the output with input λ, the empty word.

(5) Find the Moore machine that duplicates the finite automaton

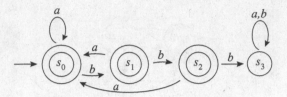

(6) Find the Moore machine that duplicates the finite automaton

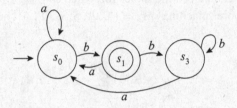

(7) Find the Moore machine that duplicates the finite automaton

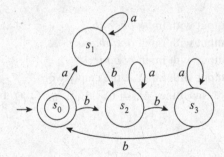

(8) Find the Moore machine that duplicates the finite automaton

(9) Let the Mealy machine $M_e = (\Sigma, A, S, s_0, \Upsilon, \delta)$ be given by the diagram

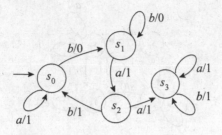

Describe A, Σ, and S. Find tables for Υ and δ.

(10) Let the Mealy machine $M_e = (\Sigma, A, S, s_0, \Upsilon, \delta)$ be given by the diagram

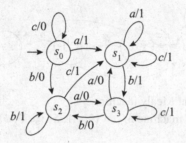

Describe A, Σ, and S. Find tables for Υ and δ.

(11) Let the Mealy machine $M_e = (\Sigma, A, S, s_0, \Upsilon, \delta)$ be given by the diagram

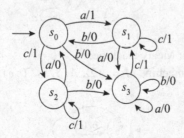

(a) Find the output with input *abaabbab*.
(b) Find the output with input *bbaaba*.
(c) Find the output with input *aabbaaa*.
(d) Find the output with input λ, the empty word.

(12) Let the Mealy machine $M_e = (\Sigma, A, S, s_0, \Upsilon, \delta)$ be given by the diagram

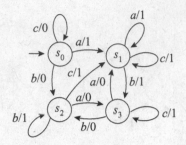

    (a) Find the output with input *abcccbab*.
    (b) Find the output with input *bbaabc*.
    (c) Find the output with input *aaccbba*.

(13) Given the Moore machine $M_o = (\Sigma, A, S, s_0, \Upsilon, \phi)$

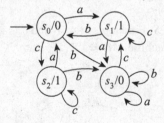

find the equivalent Mealy machine.

(14) Given the Moore machine $M_o = (\Sigma, A, S, s_0, \Upsilon, \phi)$

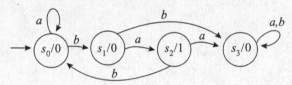

find the equivalent Mealy machine.

(15) Given the Mealy machine $M_e = (\Sigma, A, S, s_0, \Upsilon, \delta)$

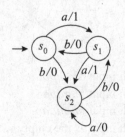

find the equivalent Moore machine.

(16) Given the Mealy machine $M_e = (\Sigma, A, S, s_0, \Upsilon, \delta)$

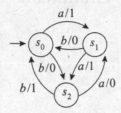

find the equivalent Moore machine.

(17) Construct a Mealy machine which directly subtracts a signed binary number from another signed binary number.

(18) Let $Z_5 = \{\bar{0}, \bar{1}, \bar{2}, \bar{3}, \bar{4}\}$ be the set of integers modulo 5, where the "sum" of two integers is found by adding the numbers and finding the remainder of this sum when divided by 5. Therefore $\bar{3} + \bar{4} = \bar{2}$ and $\bar{2} + \bar{3} = \bar{0}$. Construct the Moore machine that gives a sum of initial strings of elements of Z_5. Thus the input $\bar{2}\bar{1}\bar{4}\bar{0}\bar{3}\bar{2}\bar{1}$ produces output $\bar{0}\bar{2}\bar{3}\bar{2}\bar{2}\bar{0}\bar{2}\bar{3}$.

4
Grammars

4.1 Formal grammars

A **grammar** is intuitively a set of rules which are used to construct a language contained in Σ^* for some alphabet Σ. These rules allow us to replace symbols or strings of symbols with other symbols or strings of symbols until we finally have strings of symbols contained in Σ allowing us to form an element of the language. By placing restrictions on the rules, we shall see that we can develop different types of languages. In particular we can restrict our rules to produce desirable qualities in our language. For example in our examples below we would not want $3 + \div 4 - \times 6$. We also would not want a sentence *Slowly cowboy the leaped sunset.* Suppose that we begin with a word *add*, and that we have a rule that allows us to replace *add* with $A + B$ and that both A and B can be replaced with any nonnegative integer less that ten. Using this rule, we can replace A with 5 and B with 3 to get $5 + 3$. There might also be an additional rule that allows us to replace *add* with a different string of symbols.

If we add further rules that A can be replaced by $A + B$ and B can be replaced by $A \times B$, we can start by replacing *add* with $A + B$. If we then replace A with $A + B$ and B with $A \times B$, we get $A + B + A \times B$. We can continue this process getting longer and longer strings, so that we can continue to build strings of arbitrary length, but eventually we will want to replace all of the As and Bs with integers. As noted above, we have choices in the replacement of A and B so there is not necessarily any uniqueness in replacing a symbol or string of symbols. Hence grammars are not deterministic. If we have derived $A + A \times B$ and choose to replace the As and Bs with integers we do not have to replace both by the same As with the same value. If we replace the first A with 3, the second A with 5, and B with 7, we have $3 + 5 \times 7$.

Note that in the above rules *add*, A, and B can be replaced by other symbols while $+$ and the integers cannot be replaced. The symbols that can be replaced by other symbols are called **nonterminal symbols** and the symbols that can not be replaced by other symbols are called **terminal symbols**. We generate an element of the language when the string consists only of terminal symbols. The rules which tell us how to replace symbols are called **productions**. We denote the production (or rule) which tells us that *add* can be replaced with $A + B$

$$add \rightarrow A + B.$$

Thus the productions for our first example above are

$$add \rightarrow A + B$$
$$A \rightarrow A + B$$
$$B \rightarrow A \times B$$

$A \rightarrow 0$	$B \rightarrow 0$
$A \rightarrow 1$	$B \rightarrow 1$
$A \rightarrow 2$	$B \rightarrow 2$
$\vdots \quad \vdots \quad \vdots$	$\vdots \quad \vdots \quad \vdots$
$A \rightarrow 9$	$B \rightarrow 9.$

Below, we shall expand our rules to do arbitrary addition, subtraction, multiplication, and division of integers.

A grammar is formally defined as follows:

Definition 4.1 *A formal grammar or phrase structure grammar Γ is denoted by the 4-tuple (N, Σ, S, P) which consists of a finite set of nonterminal symbols N, a finite set of terminal symbols Σ, an element $S \in N$, called the start symbol and a finite set of productions P, which is a relation in $(N \cup \Sigma)^*$ such that each first element in an ordered pair of P contains a symbol from N and at least one production has S as the left string in some ordered pair.*

Definition 4.2 *If W and W' are elements of $(N \cup \Sigma)^*$, $W = uvw$, $W' = uv'w$, and $v \rightarrow v'$ is a production, this is denoted by $W \Rightarrow W'$. If*

$$W_1 \Rightarrow W_2 \Rightarrow W_3 \Rightarrow \cdots \Rightarrow W_n$$

*for $n \geq 1$, then W_n is derived from W_1. This is denoted by $W_1 \Rightarrow_n^* W_n$ and is called a **derivation**. If the number of productions in not important we simply use $W_1 \Rightarrow^* W_n$. The set of all strings of elements of Σ which may be generated by the set of productions P is called the **language generated by the grammar** Γ and is denoted by $\Gamma(L)$.*

To generate a word from the grammar Γ, we keep using productions to derive new strings until we have a string consisting only of terminal elements.

Thus in our example above,

$$N = \{add, A, B\},$$

$$\Sigma = \{+, \times, 0, 1, 2, 3, 4, 5, 6, 7, 8, 9\},$$

$$S = add,$$

and

$$P = \{(add, A + B), (A, A + B), (B, A \times B), (A, 0),$$

$$(A, 1), \ldots, (A, 9), (B, 0), (B, 1), \ldots, (B, 9)\},$$

where we will denote $(add, A + B)$ by $add \to A + B$, $(A, A + B)$ by $A \to A + B$, etc. If we eliminate the production $(B, A \times B)$, the language generated by Γ is the set of all formal expressions of finite sums of nonnegative integers less than 10.

Example 4.1 In the grammar described above, derive the expression

$$2 + 4 + 7 \times 6.$$

Begin with the production

$$add \to A + B$$

to derive

$$A + B.$$

Then use the production

$$B \to A \times B$$

to derive

$$A + A \times B.$$

Then use the production

$$A \to A + B$$

to derive

$$A + A + B \times B.$$

Then use the productions

$$A \to 2 \quad A \to 4 \quad B \to 7 \quad B \to 6$$

to derive $2 + 4 + 7 \times 6$. Note that we cannot derive

$$3 \times 2 + 4 + 7 \times 6.$$

Example 4.2 Suppose we want a grammar which derives arithmetic expressions for the set of integers $\{0, 1, 2, 3, 4, 5, 6, 7, 8, 9\}$. Thus the language generated by the grammar is the set of all finite arithmetic expressions for the set of integers $\{0, 1, 2, 3, 4, 5, 6, 7, 8, 9\}$. Examples would be $3 \times (5 + 4)$ and $(4 + 5) \div (3\char`^2)$, where $\char`^$ denotes exponent. As mentioned above, we obviously want to exclude expressions such as $3 + \times 6$ and $3 + \div 6 \times 4 - 5$. Let the set $N = \{S, A, B\}$ and $\Sigma = \{+, -, \times, \div, \char`^, 0, 1, 2, 3, 4, 5, 6, 7, 8, 9, (,)\}$. We will need the following productions:

$$
\begin{array}{ll}
S \to (A + B) & B \to (A + B) \\
S \to (A - B) & B \to (A - B) \\
S \to (A \times B) & B \to (A \times B) \\
S \to (A \div B) & B \to (A \div B) \\
S \to (A\char`^B) & B \to (A\char`^B) \\
A \to (A + B) & A \to 0 \\
A \to (A - B) & \vdots \quad \vdots \quad \vdots \\
A \to (A \times B) & A \to 9 \\
A \to (A \div B) & B \to 0 \\
A \to (A\char`^B) & \vdots \quad \vdots \quad \vdots \\
& B \to 9.
\end{array}
$$

We will use the grammar to derive the arithmetic expression

$$((2 + 3) \div (4 + 5)).$$

We begin with the production

$$S \to (A \div B).$$

We then use the productions

$$A \to (A + B)$$

and

$$B \to (A + B)$$

to derive

$$((A + B) \div (A + B)).$$

The productions

$$A \to 2 \quad \text{and} \quad B \to 3$$

give us

$$((2 + 3) \div (A + B)).$$

Finally we use the productions

$$A \to 4 \quad \text{and} \quad B \to 5$$

to derive

$$((2 + 3) \div (4 + 5)).$$

We next use the grammar to derive the arithmetic expression

$$((3\char`^2) \div (5 \times 7)).$$

We begin with the production

$$S \to (A \div B).$$

We then use the productions

$$A \to (A\char`^B) \quad \text{and} \quad B \to (A \times B)$$

to derive

$$((A\char`^B) \div (A \times B)).$$

The productions

$$A \to 3 \quad \text{and} \quad B \to 2$$

give us

$$((3\char`^2) \div (A \times B)).$$

Finally we use the productions

$$A \to 5 \quad \text{and} \quad B \to 7$$

to derive

$$((3\char`^2) \div (5 \times 7)).$$

Example 4.3   In a similar manner, we may form arithmetic expressions in postfix notation. Let the set $N = \{S, A, B\}$ and

$$\Sigma = \{+, -, \times, /, \char`^, 0, 1, 2, 3, 4, 5, 6, 7, 8, 9\}.$$

We will need the following productions:

$$S \to AB+ \qquad A \to AB+ \qquad B \to AB+ \qquad A \to 0$$
$$S \to AB- \qquad A \to AB- \qquad B \to AB- \qquad \vdots$$
$$S \to AB\times \qquad A \to AB\times \qquad B \to AB\times \qquad A \to 9$$
$$S \to AB\div \qquad A \to AB\div \qquad B \to AB\div \qquad B \to 0$$
$$S \to AB^\wedge \qquad A \to AB^\wedge \qquad B \to AB^\wedge \qquad \vdots$$
$$B \to 9.$$

Consider the expression 3 2 + 4 7 + ×. Since our integers are all less than ten, 3 2+ represents the integer symbol 3, followed by the integer symbol 2 and the + symbol. To construct this expression we begin with the production

$$S \to AB \times .$$

We then use the productions

$$A \to A + B \quad \text{and} \quad B \to A + B$$

to derive

$$AB + AB + \times.$$

The productions

$$A \to 2 \quad \text{and} \quad B \to 3$$

give us

$$2 \quad 3 + AB + \times.$$

Finally we use the productions

$$A \to 4 \quad \text{and} \quad B \to 7$$

to derive

$$2 \quad 3 + 4 \quad 7 + \times.$$

Example 4.4 A grammar may also be used to derive proper sentences. These sentences are proper in the sense that they are grammatically correct, although they may not have any meaning. Suppose we want a grammar which will derive the following statements, among others:

> Joe chased the dog.
> The fast horse leaped over the old fence.
> The cowboy rode slowly into the sunset.

Before actually stating the grammar let us decide upon its structure. This allows us to be assured that each sentence in the grammar is a grammatically correct sentence. Each of our sentences has a noun phrase (noun p), a verb phrase (verb p), and another noun phrase. In addition the last two sentences have a preposition (prep). Therefore let the first production be

$$S \rightarrow\ <\text{noun p}><\text{verb p}><\text{prep}><\text{noun p}>.$$

In our example, the most general form of a noun phrase is an article followed by an adjective and then a noun. Therefore let the next production be

$$<\text{noun phrase}>\ \rightarrow\ <\text{art}><\text{adj}><\text{noun}>$$

where "art" represents article and "adj" represents adjective

The most general form of a verb phrase is a verb followed by an adverb. Therefore let the next production be

$$<\text{verb p}>\ \rightarrow\ <\text{adv}><\text{verb}>$$

where "adv" represents adverb.

At this point, we know that the terminal set $\Sigma = \{$Joe, chased, the, The, dog, fast, horse, leaped, over, old, fence, cowboy, rode, slowly, into, sunset$\}$. The nonterminal set $N = \{S$, <noun p>, <verb p>, <art>, <adj>, <noun>, <adv>, <verb>, <prep>$\}$.

We next need productions which will assign values to <art>, <adj>, <noun>, <adv>, and <verb>. In some of our sentences we do not need <art>, <adjective>, <prep>, and <adv>. To solve this problem, we include the productions

$$<\text{art}>\rightarrow\lambda \qquad <\text{adj}>\rightarrow\lambda \qquad <\text{adv}>\rightarrow\lambda \qquad <\text{prep}>\rightarrow\lambda.$$

By assigning these symbols to the empty set, we simply erase them when they are not needed. The remainder of our productions consists of the following:

< art > → the	< noun > → horse	< noun > → fence
< adj > → fast	< noun > → dog	< adv > → slowly
< adj > → old	< noun > → cowboy	< verb > → chased
< noun > → Joe	< noun > → sunset	< verb > → leaped
< verb > → rode	< prep > → over	< prep > → into
< art > → The.		

To derive the sentence "Joe chased the dog," we begin with

$$S \rightarrow\ <\text{noun p}><\text{verb p}><\text{prep}><\text{noun p}>$$

to derive

$$< \text{noun p} > < \text{verb p} > < \text{prep} > < \text{noun p} > .$$

Using the production

$$< \text{noun p} > \rightarrow < \text{article} > < \text{adjective} > < \text{noun} >$$

we derive

$$< \text{art} > < \text{adj} > < \text{noun} > < \text{verb p} > < \text{prep} > < \text{noun p} > .$$

Using

$$< \text{art} > \rightarrow \lambda \qquad\qquad < \text{adj} > \rightarrow \lambda$$

we derive

$$< \text{noun} > < \text{verb p} > < \text{prep} > < \text{noun p} > .$$

Repeating the process for the second <noun phrase>, we derive

$$< \text{noun} > < \text{verb p} > < \text{prep} > < \text{art} > < \text{noun} > .$$

Using

$$< \text{verb p} > \rightarrow < \text{adv} > < \text{verb} > ,$$

we derive

$$< \text{noun} > < \text{adv} > < \text{verb} > < \text{prep} > < \text{art} > < \text{noun} > .$$

Using

$$< \text{adv} > \rightarrow \lambda \qquad < \text{prep} > \rightarrow \lambda$$

we derive

$$< \text{noun} > < \text{verb} > < \text{art} > < \text{noun} > .$$

Using

$$< \text{noun} > \rightarrow \text{Joe} \quad < \text{noun} > \rightarrow \text{dog} \quad < \text{verb} > \rightarrow \text{chased} \quad < \text{art} > \rightarrow \text{the}$$

we derive "Joe chased the dog."

To derive the sentence "The fast horse leaped over the tall fence," we again begin with

$$S \rightarrow < \text{noun p} > < \text{verb p} > < \text{prep} > < \text{noun p} >$$

to derive

$$< \text{noun p} > < \text{verb p} > < \text{prep} > < \text{noun p} > .$$

Using the production

$$< noun\ p > \rightarrow < art >< adj >< noun >$$

we derive

$$< art >< adj >< noun >< verb\ p >< prep >< noun\ p > .$$

Using

$$< art > \rightarrow the \quad < art > \rightarrow The \quad < adj > \rightarrow fast \quad < noun > \rightarrow horse$$

we derive

$$The\ fast\ horse < verb\ p >< prep >< noun\ p > .$$

Using

$$< verb\ p > \rightarrow < adv >< verb >,$$

we derive

$$The\ fast\ horse < adv >< verb >< prep >< noun\ p > .$$

Using

$$< adv > \rightarrow \lambda \qquad < verb > \rightarrow leaped$$

we derive

$$The\ fast\ horse\ leaped\ < prep >< noun\ p > .$$

Using

$$< prep > \rightarrow over,$$

we derive

$$The\ fast\ horse\ leaped\ over < noun\ p > .$$

Using the production

$$< noun\ p > \rightarrow < art >< adj >< noun >$$

we derive

$$The\ fast\ horse\ leaped\ over < art >< adj >< noun > .$$

Using

$$< art > \rightarrow the \qquad < adj > \rightarrow tall \qquad < noun > \rightarrow fence$$

we derive

The fast horse leaped over the tall fence.

Derivation of the last sentence is left to the reader.

Definition 4.3 *For each production $P \to w_1 w_2 w_3 \ldots w_n$ the **corresponding** tree is*

Thus the corresponding tree for $S \to A + B$ is

Definition 4.4 *If the corresponding trees of the productions used to derive a given expression are connected, they form a tree with root S, called the **parse tree** or the **derivation tree**. If $A \to B$ occurs in the derivation then there is an **edge** from A to B in the tree. The symbols A and B are called **vertices** or **nodes**. The vertex B is called the **child** of A. Note that a terminal at a vertex has no children. Such a vertex is called a **leaf** of the tree. The leaves of the tree, when read left to right, form the word generated by the tree. If $A_0 \to A_1 \to \cdots \to A_n$ forms a string of edges in the tree then there is a **path** of length n from A_0 to A_n.*

Example 4.5 In Example 4.1, we used productions to derive $3 + 2 + 4$.

To construct the tree, begin with the first production used

$$\text{add} \to A + B$$

to form corresponding tree

Then use the corresponding tree

of the production

$$A \rightarrow A + B$$

to form the tree

Then use the corresponding tree in

of the production

$$B \rightarrow A \times B$$

to get the corresponding tree

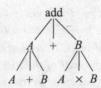

Then use the corresponding trees of the next productions

$$A \rightarrow 2 \qquad B \rightarrow 4 \qquad A \rightarrow 7 \qquad B \rightarrow 6$$

to form the parse tree

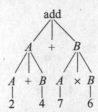

Example 4.6 In Example 4.3, to derive $((2+3) \times (4+5))$, we use the productions

$$S \rightarrow (A \times B) \qquad A \rightarrow (A + B) \qquad A \rightarrow 2 \qquad B \rightarrow 3$$
$$B \rightarrow (A + B) \qquad A \rightarrow 4 \qquad B \rightarrow 5.$$

Therefore the parse tree is the tree

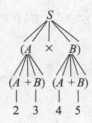

Example 4.7 In Example 4.4, to derive the sentence "Joe chased the dog," using productions

$$S \to < \text{noun p} >< \text{verb p} >< \text{prep} >< \text{noun p} >$$

to get

$$< \text{noun p} >< \text{verb p} >< \text{prep} >< \text{noun p} >$$

$$< \text{noun p} > \to < \text{article} >< \text{adjective} >< \text{noun} >$$

to get

$$< \text{art} >< \text{adj} >< \text{noun} >< \text{verb p} >< \text{prep} >< \text{noun p} >$$

Using

$$< \text{art} > \to \lambda \qquad < \text{adj} > \to \lambda$$

we get

$$< \text{noun} >< \text{verb p} >< \text{prep} >< \text{noun p} >$$

Again using

$$< \text{noun p} > \to < \text{article} >< \text{adjective} >< \text{noun} >$$

and

$$< \text{art} > \to \lambda \qquad < \text{adj} > \to \lambda$$

we get

$$< \text{noun} ><.\text{verb p} >< \text{prep} >< \text{art} >< \text{noun} >.$$

Using

$$< \text{verb p} > \to < \text{adv} >< \text{verb} >,$$

we get

$$< noun >< adv >< verb >< prep >< art >< noun >.$$

Using

$$< adv > \rightarrow \lambda \qquad < prep > \rightarrow \lambda$$

we get

$$< noun >< verb >< art >< noun >.$$

Using

$$< noun > \rightarrow Joe \quad < noun > \rightarrow dog \quad < verb > \rightarrow chased \quad art \rightarrow the$$

we have the correspondence tree for "Joe chased the dog."

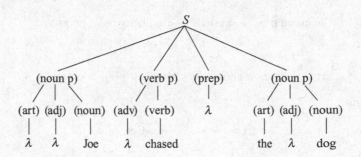

Example 4.8 In Example 4.4, to derive the sentence "The large dog leaped over the old fence," we use productions

$$S \rightarrow < noun\ p >< verb\ p >< prep >< noun\ p >$$

$$< noun\ p > \rightarrow < art >< adj >< noun >$$

$< noun\ p > \rightarrow < art >< adj >< noun >$	$< adj > \rightarrow fast$
$< verb\ p > \rightarrow < adv >< verb >$	$< adv > \rightarrow \lambda$
$< prep > over$	$< art > \rightarrow The$
$< adj > \rightarrow tall$	$< noun > \rightarrow fence$
$< noun > \rightarrow horse$	$< verb > \rightarrow leaped$
$< art > \rightarrow the.$	

Thus the parse tree is

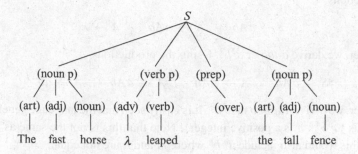

In all of the grammars in this section, the productions have been of the form $A \to W$, where A is a nonterminal symbol. Therefore the production can be used everywhere that A appears, regardless of its position in an expression. Such grammars are called **context-free grammars**. A language generated by a context-free grammar is called a **context-free language.** If a grammar has a production of the form $aAb \to W$ where A is a nonterminal and $ab \neq \lambda$ then this production can only be used when a is on the left-hand side of A and b is on the right-hand side. It therefore cannot be used whenever A appears and so it is dependent on the context in which A appears. Such a grammar is called a **context-sensitive grammar**.

In the following examples, we consider context-free grammars which generate more abstract languages:

Example 4.9 Let $\Gamma = (N, \Sigma, S, P)$ be the grammar defined by $N = \{S, A, B\}$, $\Sigma = \{a, b\}$ and P be the set of productions

$$S \to AB \qquad A \to a \qquad B \to Bb \qquad B \to \lambda \qquad A \to \lambda \qquad A \to aA.$$

Using the production $S \to AB$, we derive AB. Next using the productions $A \to a$ and $B \to \lambda$, we derive a. If we use the productions

$$S \to AB \qquad A \to \lambda \qquad B \to Bb \qquad B \to \lambda$$

in order, we derive b. We can also generate $aabbb$, $aaaa$, $aaab$, and $bbbbb$. In fact, we can generate $a^m b^n$ for all nonnegative integers m, n. Hence the expression for the language generated by Γ is a^*b^*.

Example 4.10 Let $\Gamma' = (N, \Sigma, S, P)$ be the grammar defined by $N = \{S, A\}$, $\Sigma = \{a, b\}$ and P be the set of productions

$$S \to aAb \qquad A \to aAb \qquad A \to \lambda.$$

Using the productions $S \to aAb$ and $A \to \lambda$ we derive ab. Using the productions

$$S \to aAb \qquad A \to aAb \qquad A \to \lambda$$

in order, we derive $aabb$ or a^2b^2. Using the productions

$$S \to aAb \qquad A \to aAb \qquad A \to aAb \qquad A \to ab$$

in order, we derive $aaabbb$ or a^3b^3. It is easily seen that the language generated by Γ' is $\{a^n b^n : n$ is a positive integer$\}$. Note that this is not the same as a^*b^* since this would also include $a^m b^n$ where m and n are not equal.

Example 4.11 Let $\Gamma'' = (N, \Sigma, S, P)$ be the grammar defined by $N = \{S, A, B\}$, $\Sigma = \{a, b\}$ and P be the set of productions

$$S \to ABABABA \qquad A \to Aa \qquad A \to \lambda \qquad B \to b.$$

It can be shown that the expression for the language generated by Γ'' is $a^*ba^*ba^*ba^*$. This is the language consisting of all words containing exactly three bs.

In Example 4.10 we generated the language $\{a^n b^n : n$ is a positive integer$\}$. Intuitively, we can see that this is not a regular language since the only way that we can generate an infinite regular language using a finite alphabet is with the Kleene star $*$. In this case the only possibility is a^*b^* but, as mentioned earlier, this does not work since this would also include $a^m b^n$ where m and n are not equal.

One might ask if there is a particular type of grammar which generates only regular languages. The answer is yes, as we shall now show.

Definition 4.5 *A context-free grammar* $\Gamma = (N, \Sigma, S, P)$ *is called a* **regular grammar** *if every production* $p \in P$ *has the form* $n \to w$ *where* w *is the empty word* λ *or the string* w *contains at most one nonterminal symbol and it occurs at the end of the string if at all.*

Therefore w could be of the form $aacA$, ab, λ or bA, where a, b, and c are terminals and A is a nonterminal. However, w could not be of the form aAb, aAB, or Aa. The production $n \to abcA$ could be replaced by the productions

$$n \to aB \qquad B \to bC \qquad C \to cA.$$

Also it is possible w could contain no terminal and one nonterminal so we have $B \to C$, but if this is followed by $C \to tD$, where t is a terminal, then we can combine the two productions to get $B \rightarrow tD$. Hence it is no restriction to require each production to be one of the following forms:

$$A \rightarrow aB \qquad B \rightarrow b \qquad C \rightarrow \lambda$$

where A, B, and C are nonterminal elements, and a and b are terminal elements.

More formally, we define a linear regular grammar as follows:

Definition 4.6 *A context-free grammar* $\Gamma = (N, \Sigma, S, P)$ *is called a* **linear regular grammar** *if every production* $p \in P$ *has the form* $n \rightarrow w$ *where the string* w *has the form* xY, x *or* λ *where* $x \in \Sigma$ *and* $Y \in N$.

Theorem 4.1 *A language is generated by a linear regular grammar if and only if it is generated by a regular grammar.*

Proof Obviously every language that is generated by a linear regular grammar is generated by a regular grammar. To show every regular grammar is generated by a linear regular grammar, we divide the proof into two parts. We first show the language of a regular grammar can be generated by productions of the forms

$$A \rightarrow aB \qquad B \rightarrow b \qquad C \rightarrow \lambda \qquad C \rightarrow D$$

where A, B, C, and D are nonterminals and a, b are terminals. Let $\Gamma = (N, \Sigma, S, P)$ be a regular grammar and L be the language generated by Γ. Let $\Gamma' = (N', \Sigma, S, P')$ be the grammar formed by replacing every production $A \rightarrow a_1 a_2 a_3 \ldots a_{n-1} B$ by the set of productions $A \rightarrow a_1 A_1$, $A_1 \rightarrow a_2 A_2, \ldots, A_{n-1} \rightarrow a_{n-2} A_{n-2}$, $A_n \rightarrow a_{n-1} B$ where $A_1, A_2, \ldots, A_n$ are new nonterminal symbols. Let L' be the language generated by Γ'. By construction we have $A \Rightarrow^* a_1 a_2 a_3 \ldots a_{n-1} B$. So any word of L will be created by the grammar Γ'. Conversely if $A \Rightarrow^* a_1 a_2 a_3 \ldots a_{n-1} B$ is formed by productions $A \rightarrow a_1 A_1$, $A_1 \rightarrow a_2 A_2, \ldots, A_{n-1} \rightarrow a_{n-2} A_{n-2}$, $A_n \rightarrow a_{n-1} B$, then there must be a production $A \rightarrow a_1 a_2 a_3 \ldots a_{n-1} B$ in Γ since the symbols $A_1, A_2, \ldots, A_n$ are symbols which appear only in forming $A \rightarrow a_1 a_2 a_3 \ldots a_{n-1} B$.

Hence we can now assume that a regular grammar can be formed using only productions of the form

$$A \rightarrow aB \qquad B \rightarrow b \qquad C \rightarrow \lambda \qquad C \rightarrow D$$

where A, B, C, and D are nonterminals and a, b are terminals. We want to show that we can form a regular grammar without productions of the form $C \rightarrow D$ where C and D are both nonterminals. Call this a 1-production. Let Γ be a regular grammar formed using the productions above and L be the language generated by Γ. Assume that we have productions of the form above. Let Γ'' be the grammar with all 1-productions deleted and insert the production

$A_1 \rightarrow A_n b$ if

$$A_1 \rightarrow A_2, A_2 \rightarrow A_3, \cdots, A_{n-2} \rightarrow A_{n-1}, A_{n-1} \rightarrow bA_n$$

occurred in L.

If

$$A_1 \rightarrow A_2, A_2 \rightarrow A_3, \cdots, A_{n-2} \rightarrow A_{n-1}, A_{n-1} \rightarrow b$$

occurred in L, insert the production $A_1 \rightarrow b$ in L''.

If

$$A_1 \rightarrow A_2, A_2 \rightarrow A_3, \cdots, A_{n-2} \rightarrow A_{n-1}, A_{n-1} \rightarrow \lambda$$

occurred in L, insert the production $A_1 \rightarrow \lambda$ in L''. Let L'' be the language generated by the grammar Γ''. Certainly $L \subseteq L''$.

Assume we have $S \Rightarrow^* w$ where the productions are from Γ'' or Γ or both and $w \in \Sigma^*$. If all of the productions are from Γ, then $w \in L$. If not then there exists $uB \Rightarrow vC$ in the sequence where $B \rightarrow aC$ is not a production of Γ and $v = ua$. Take the first such production. Therefore there exist productions $B \rightarrow A_1, A_1 \rightarrow A_2, A_2 \rightarrow A_3, \cdots, A_{n-2} \rightarrow A_{n-1}, A_{n-1} \rightarrow aC$ in Γ and we can replace $uB \Rightarrow vC$ with $uB \Rightarrow uA_1 \Rightarrow uA_2 \cdots uA_{n-1} \Rightarrow uaC = vC$, where all of the productions are in Γ. Since there are only a finite number of productions not in Γ, we can continue this process until all of the productions are in Γ and $w \in L$. Therefore $L' \subseteq L''$. □

We now proceed to prove the following theorem.

Theorem 4.2 *A language is regular if and only if it is generated by a regular grammar.*

Since a language is regular if and only if it is accepted by an automaton, all we need to know is that a language is generated by a regular grammar if and only if it is accepted by an automaton. We first show how to construct a regular grammar which generates the same language that is accepted by a given deterministic automaton and we then show how to construct an automaton which accepts the language generated by a given regular grammar.

Normally when we consider a word being read by an automaton, we probably think of the automaton as removing letters from the word as it reads it. Thus if the word to be read is $abbc$, and there is an a-arrow from state s_0 to s_1, then we read a, move to state s_1, and still have bbc left to read. If there is a b-arrow from state s_1 to s_2 then we read b, move to state s_2, and still have bc left to read. If there is a b-arrow from state s_2 to s_3 then we read b, move to state s_3, and still have c left to read. Finally, if there is a c-arrow from state s_3 to s_4 then we read c, move to state s_4, and have nothing left to read. If s_4 is a terminal state, then we accept the word $abbc$.

We may also think of an automaton adding letters to words rather than removing them. Suppose that we consider the string that has been read rather than the string left to read. In the example above, at state s_1, we have read a. At state s_2 we have read ab. At state s_3 we have read abb, and at state s_4 we have read $abbc$. Thus at each state we are adding a letter. Consider the grammar $\Gamma = (N, \Sigma, s_0, P)$, where $N = \{s_0, s_1, s_2, s_3, s_4\}$, $\Sigma = \{a, b, c\}$, and P is the set of productions

$$s_0 \to as_1 \qquad s_1 \to bs_2 \qquad s_2 \to bs_3 \qquad s_3 \to cs_4 \qquad s_4 \to \lambda$$

where we have a production $s_4 \to \lambda$ only if s_4 is a terminal state. It is easily seen that Γ generates the word $abbc$. Thus to change an automaton to a regular grammar, if there is a k-arrow from s_i to s_j, in the corresponding grammar, form the production $s_i \to ks_j$. If s_j is an acceptance state, add the production $s_j \to \lambda$. We shall shortly show that grammar will generate the same language accepted by the automaton.

Example 4.12 Given the automaton,

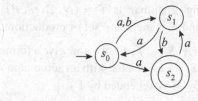

we form the productions for the corresponding grammar as follows:

Description of automaton	Production
There is an a-arrow from s_0 to s_1	$s_0 \to as_1$
There is a b-arrow from s_0 to s_1	$s_0 \to bs_1$
There is an a-arrow from s_1 to s_0	$s_1 \to as_0$
There is a b-arrow from s_1 to s_2	$s_1 \to bs_2$
There is an a-arrow from s_2 to s_1	$s_2 \to as_1$
There is a b-arrow from s_0 to s_2	$s_0 \to bs_2$
The state s_2 is an acceptance state	$s_2 \to \lambda$.

Hence the corresponding grammar is $\Gamma = (N, \Sigma, s_0, P)$, where $N = \{s_0, s_1, s_2\}$, $\Sigma = \{a, b\}$, and P is the above set of productions.

Example 4.13   Given the automaton

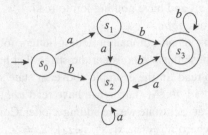

we form the productions for the corresponding grammar as follows:

Description of automaton	Production
There is an a-arrow from s_0 to s_1	$s_0 \rightarrow as_1$
There is a b-arrow from s_0 to s_2	$s_0 \rightarrow bs_2$
There is an a-arrow from s_1 to s_2	$s_1 \rightarrow as_2$
There is a b-arrow from s_1 to s_3	$s_1 \rightarrow bs_3$
There is an a-arrow from s_2 to s_2	$s_2 \rightarrow as_2$
There is a b-arrow from s_2 to s_3	$s_2 \rightarrow bs_3$
There is an a-arrow from s_3 to s_2	$s_3 \rightarrow as_2$
There is a b-arrow from s_3 to s_3	$s_3 \rightarrow bs_3$
The state s_2 is an acceptance state	$s_2 \rightarrow \lambda$
The state s_3 is an acceptance state	$s_3 \rightarrow \lambda$.

Hence the corresponding grammar is $\Gamma = (N, \Sigma, s_0, P)$, where $N = \{s_0, s_1, s_2, s_3\}$, $T = \{a, b\}$, and P is the above set of productions.

Given an automata $M = (A, S, s_0, T, F)$, we now give a formal definition of the grammar $\Gamma_M = (N, \Sigma, S, P)$, associated with an automaton and then show that the language accepted by M is generated by Γ_M.

Definition 4.7 $\Gamma_M = (N, T, S, P)$, *the **grammar associated with the automaton** $M = (\Sigma, Q, s_0, T, F)$ has $N = Q$, and $s_0 = S$. The production $s_i \rightarrow as_j$ is in P if and only if $F(a, s_i) = s_j$, and $s_j \rightarrow \lambda$ if and only if s_j is an acceptance state.*

Lemma 4.1   *The language L_1 accepted by M is equal to the language L_2 generated by Γ_M.*

Proof   From the above definition, we have $s_i \rightarrow as_j$ if and only if $(s_i, a) \vdash (s_j, \lambda)$. Thus if $(s_i, ab) \vdash (s_j, b) \vdash (s_k, \lambda)$ in M then $s_i \Rightarrow as_j \Rightarrow abs_k$ in Γ_M. More generally, $(s_i, w) \vdash^* (s_k, \lambda)$ if and only if $s_i \Rightarrow^* ws_k$.

We first show $L_1 \subseteq L_2$. Assume w is accepted by M, then $(s_0, w) \vdash^* (s_k, \lambda)$ where s_k is an acceptance state. Since $(s_0, w) \vdash^* (s_k, \lambda)$, we have $s_0 \Rightarrow^* ws_k$ in Γ_M. Since s_k is an acceptance state, $s_k \rightarrow \lambda$ is a production. Therefore $s_0 \Rightarrow^* w$ and w is generated by Γ_M.

Conversely, let w be generated by Γ_M. Let $s_0 \Rightarrow^* w$, then $s_0 \Rightarrow^* w s_k \Rightarrow w$. Therefore $s_k \to \lambda$ is a production, and by definition of Γ_M, s_k is an acceptance state. Since $s_0 \Rightarrow^* w$, we have $(s_0, w) \vdash^* (s_k, \lambda)$ in M. Therefore w is accepted by M. $\square$

Given a regular grammar Γ in linear regular grammar form, we now construct an automaton which accepts this linear grammar. Given $\Gamma = (N, \Sigma, S, P)$, intuitively we add an additional nonterminal t to N and for each production $B \to a$, where a is a terminal, we remove this production and replace it with the productions $B \to at$ and $t \to \lambda$. Obviously this does not change the language of the grammar. Let $M = (\Sigma, Q, s_0, T, F)$ be the automaton in which Q is the set of nonterminals together with the additional nonterminal t, $s_0 = S$. The set F is defined by $F(a, A) = B$ if and only if $A \to aB$ is in P. The state $B \in T$ if $B \to \lambda$.

Example 4.14 Let $\Gamma = (N, \Sigma, S, P)$ be the grammar defined by $N = \{S, A, B, C\}$, $\Sigma = \{a, b, c\}$, and P be the set of productions

$$
\begin{array}{llll}
S \to aA & A \to aA & S \to bB & B \to bB \\
A \to cC & C \to cC & B \to aA & C \to \lambda.
\end{array}
$$

The corresponding automaton is

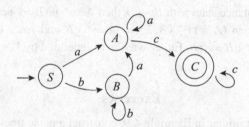

Example 4.15 Let $\Gamma = (N, T, S, P)$ be the grammar defined by $N = \{S, A, B, C\}$, $T = \{a, b, c\}$, and P be the set of productions

$$
\begin{array}{lllll}
S \to aA & A \to bB & S \to bB & B \to cC & A \to aC \\
C \to cA & B \to aA & C \to \lambda & B \to \lambda.
\end{array}
$$

The corresponding automaton is

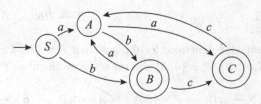

More formally, given a regular grammar $\Gamma = (N, T, S, P)$ in linear regular grammar form, we define a nondeterministic automaton M_Γ which accepts this linear grammar. Given $\Gamma = (N, T, S, P)$, let $M = (\Sigma, Q, s_0, \Upsilon, F)$ be the nondeterministic automaton in which $\Sigma = T$ and N is the set of nonterminals together with an additional nonterminal t, $s_0 = S$. The set Υ is defined by $B \in \Upsilon(a, A)$ if $A \to aB$ is in P and $t \in \Upsilon(a, A)$ if $A \to a$ is in P. The state $B \in T$ if $B \to \lambda$ or $B = t$. Hence $(A, a) \vdash (B, \lambda)$ if and only if $A \to aB$. Thus if $A \Rightarrow aB \Rightarrow abC$ in Γ then $(A, ab) \vdash (B, b) \vdash (C, \lambda)$ in M_Γ. More generally, $A \Rightarrow^* wB$ if and only if $(A, w) \vdash^* (B, \lambda)$ for nonterminals A and B.

Theorem 4.3 *The language L_1 accepted by M_Γ is equal to the language L_2 generated by Γ.*

Proof We first show $L_2 \subseteq L_1$. Let $w = va$ be generated by Γ. If $A \Rightarrow^* vaB \Rightarrow va$, then $(A, va) \vdash^* (B, \lambda)$ and since the last production is $B \to \lambda$, B is an acceptance state. If $A \Rightarrow^* vB \Rightarrow va$, then $(A, v) \vdash^* (B, \lambda)$ so $(A, va) \vdash^* (B, a)$ since the last production is $B \to a$, $(B, a) \vdash (t, \lambda)$, and t is an acceptance state. Therefore w is accepted by M_Γ.

To show $L_1 \subseteq L_2$ let $w = va$ be accepted by M_Γ, then if $(A, va) \vdash^* (B, \lambda)$ and B is an acceptance state with $B \to \lambda$ then $A \Rightarrow^* vaB \Rightarrow va$. If $(A, va) \vdash^* (B, a) \vdash (t, \lambda)$, then $(A, v) \vdash^* (B, \lambda)$ so $A \Rightarrow^* vB$ and since $(B, a) \vdash (t, \lambda)$, $B \to a$, so $A \Rightarrow^* vB \Rightarrow va$. Either way, w is generated by Γ. $\square$

Exercises

(1) Using the grammar in Example 4.9, construct a parse tree for *abbb*.

(2) Using the grammar in Example 4.10, construct a parse tree for *aaabbb*.

(3) Using the grammar in Example 4.11, construct a parse tree for *babaab*.

(4) In Example 4.4, derive the statement "The cowboy rode slowly into the sunset" and construct the correspondence parse tree.

(5) Find the language generated by the grammar $\Gamma = (N, T, S, P)$ defined by $N = \{S, A, B\}$, $T = \{a, b\}$ and the set of productions P given by

$$S \to AB \qquad A \to aA \qquad A \to \lambda \qquad B \to Bb \qquad B \to \lambda.$$

(6) Find the language generated by the grammar $\Gamma = (N, T, S, P)$ defined by $N = \{S, A, B\}$, $T = \{a, b\}$ and the set of productions P given by

$$S \to aB \qquad B \to bA \qquad A \to aB \qquad B \to b.$$

(7) Find the language generated by the grammar $\Gamma = (N, T, S, P)$ defined by $N = \{S, A, B\}$, $T = \{a, b\}$ and the set of productions P given by

$$
\begin{array}{llll}
S \to aA & B \to aA & S \to bB & A \to aB \\
B \to bB & A \to bA & B \to b & A \to a.
\end{array}
$$

(8) Find the language generated by the grammar $\Gamma = (N, T, S, P)$ defined by $N = \{S, A, B, C\}$, $T = \{a, b\}$ and the set of productions P given by

$$
\begin{array}{llll}
S \to C & A \to aB & C \to bC & B \to bB \\
C \to aA & B \to aA & A \to bA & B \to \lambda.
\end{array}
$$

(9) Find the grammar which generates the language ww^r where w is a string of as and bs and w^r is the reverse string. For example, $abba$, $abaaba$, and $abbbba$ belong to ww^r.

(10) Construct a grammar which generates the language wcw^r where $w \in \{a, b\}$ and w^r is the reverse string.

(11) Construct a grammar which generates the language $L = \{w$: where $w \in \{a, b\}$ and $w = w^r\}$.

(12) Construct a grammar which generates the language L described by the expression **aa*bb***.

(13) Construct a grammar which generates the language L described by the expression **(abc)***.

(14) Construct a grammar which generates the language L described by the expression **(ab)* $\vee$ (ac)***.

(15) Construct a grammar which generates the language L described by the expression **ac(bc)*d**.

(16) Construct a grammar which generates the language expressed by **(a*ba*ba*b)***.

(17) Construct a grammar which generates the language expressed by **(a*(ba)*bb*a)***.

(18) Construct a grammar which generates the language expressed by **(a*b) $\vee$ (b*a)***.

(19) Construct a grammar which generates the language expressed by **aa*bb*aa***.

(20) Construct a grammar which generates the language expressed by **(a*b) $\vee$ (c*b) $\vee$ (ac)***.

(21) Construct a grammar which generates the language expressed by **(a$\vee$b)*(aa $\vee$ bb)(a $\vee$ b)***.

(22) Construct a grammar which generates the language expressed by **((aa*b) $\vee$ bb*a)ac***.

(23) Construct a grammar to generate arithmetic expressions for positive integers less than ten in prefix notation.

(24) Find an automaton which accepts the language generated by the grammar $\Gamma = (N, T, S, P)$ defined by $N = \{S, A, B\}$, $T = \{a, b\}$ and the set of productions P given by

$$S \rightarrow aB \qquad B \rightarrow bA \qquad A \rightarrow aB \qquad B \rightarrow b.$$

(25) Find an automaton which accepts the language generated by the grammar $\Gamma = (N, T, S, P)$ defined by $N = \{S, A, B\}$, $T = \{a, b\}$ and the set of productions P given by

$$S \rightarrow aA \qquad B \rightarrow aA \qquad S \rightarrow bB \qquad A \rightarrow aB$$
$$B \rightarrow bB \qquad A \rightarrow bA \qquad B \rightarrow b \qquad A \rightarrow a.$$

(26) Find an automaton which accepts the language generated by the grammar $\Gamma = (N, T, S, P)$ defined by $N = \{S, A, B, C\}$, $T = \{a, b\}$ and the set of productions P given by

$$S \rightarrow C \qquad A \rightarrow aB \qquad C \rightarrow bC \qquad B \rightarrow bB$$
$$C \rightarrow aA \qquad B \rightarrow aA \qquad A \rightarrow bA \qquad B \rightarrow \lambda.$$

(27) Find an automaton which accepts the language generated by the grammar $\Gamma = (N, T, S, P)$ defined by $N = \{S, A, B, C\}$, $T = \{a, b\}$ and the set of productions P given by

$$S \rightarrow C \qquad C \rightarrow b \qquad C \rightarrow aA \qquad A \rightarrow aA$$
$$C \rightarrow aC \qquad A \rightarrow a \qquad C \rightarrow a \qquad A \rightarrow \lambda.$$

(28) Find an automaton which accepts the language generated by the grammar $\Gamma = (N, T, S, P)$ defined by $N = \{S, A, B, C\}$, $T = \{a, b\}$ and the set of productions P given by

$$S \rightarrow C \qquad C \rightarrow aaC \qquad C \rightarrow abC$$
$$C \rightarrow baC \qquad C \rightarrow bbC \qquad C \rightarrow \lambda.$$

(29) Find an automaton which accepts the language generated by the grammar $\Gamma = (N, T, S, P)$ defined by $N = \{S, A, B, C\}$, $T = \{a, b\}$ and the set of productions P given by

$$S \rightarrow C \qquad B \rightarrow aB \qquad C \rightarrow bC \qquad B \rightarrow bB \qquad C \rightarrow aA$$
$$B \rightarrow a \qquad A \rightarrow bC \qquad B \rightarrow b \qquad A \rightarrow aB.$$

(30) Construct a grammar which generates the language accepted by the automaton

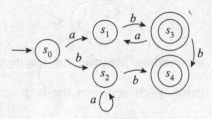

(31) Construct a grammar which generates the language accepted by the automaton

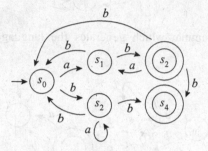

(32) Construct a grammar which generates the language accepted by the automaton

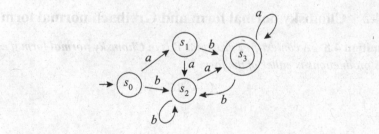

(33) Construct a grammar which generates the language accepted by the automaton

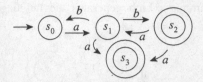

(34) Construct a grammar which generates the language accepted by the automaton
(35) Construct a grammar which generates the language accepted by the automaton

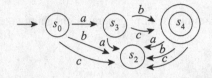

(36) Construct a grammar which generates the language accepted by the automaton

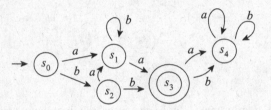

4.2   Chomsky normal form and Greibach normal form

Definition 4.8   *A context-free grammar Γ is in **Chomsky normal form** if each of its productions is either of the form*

$$A \to BC$$

or

$$A \to a$$

where A, B, and C are nonterminals and a is a terminal.

Definition 4.9   *A context-free grammar Γ is in **Greibach normal form** if each of its productions is of the form*

$$A \to aW$$

where a is a terminal and W is a possibly empty string of nonterminals.

We shall show that every language L, not containing the empty word, which is generated by a context-free grammar can be generated by a context-free grammar in Chomsky normal form. We shall also show that every language L, not containing the empty word, which is generated by a context-free grammar can be generated by a context-free grammar in Greibach normal form.

We shall first show that a language L, not containing the empty word, generated by a context-free grammar, can be generated by a context-free grammar in Chomsky normal form. We begin with a series of lemmas. The first lemma, which demonstrates the flexibility of derivations in context-free languages shows us that if we have a derivation $UV \Rightarrow^* W$ where $U, V, W \in (N \cup T)^*$ then U and V may be treated separately.

Lemma 4.2 *Let $\Gamma = (N, T, S, P)$ be a grammar and $UV \Rightarrow^* W$, where $U, V, W \in (N \cup T)^*$, be a derivation in Γ with n steps, then W can be expressed as $W_1 W_2$ where $U \Rightarrow^* W_1$, $V \Rightarrow^* W_2$ are derivations in Γ, both containing at most n steps.*

Proof   The proof of this lemma uses induction on the number of steps in the production. Assume there is one step. Then only one nonterminal is replaced using a production. Assume it is the production $A \rightarrow wBw'$. Either A is in the string U or is in the string V. Without loss of generality assume it is in U, so

$$U = XAY$$

and

$$U \Rightarrow XwBw'Y = U'.$$

Further

$$UV \Rightarrow XwBw'YV = U'V,$$

so letting $W_1 = U'$ and $W_2 = V$ we are done.

Assume the lemma is true for all derivations with less than k steps. Assume $UV \Rightarrow^* W$ contains k steps. As above assume the first step is $UV \Rightarrow U'V$ where $U \Rightarrow U'$. Note that $U'V \Rightarrow^* W$ uses only $k - 1$ steps. By induction there are derivations $U' \Rightarrow^* W_1$, $V \Rightarrow^* W_2$ containing at most k steps. Therefore $U \Rightarrow U', U' \Rightarrow^* W_1, V \Rightarrow^* W_2$ are the required derivations. $\square$

One of the results of this lemma is that we can get from $UV \Rightarrow^* W$ by the derivations

$$UV \Rightarrow^* W_1 V \Rightarrow^* W_1 W_2$$

where $W = W_1 W_2$ since if $X_\alpha \Rightarrow X_\beta$ is a derivation, then so are $X_\alpha V \Rightarrow X_\beta V$ and $UX_\alpha \Rightarrow UX_\beta$.

The next lemma shows us that in applying productions, the order in which we apply them is not particularly important. In fact we can derive any word w in the language generated by Γ by replacing the leftmost nonterminal by a production at each step in the derivation. This is called a **leftmost derivation** of w.

Lemma 4.3 *Let* $w \in \Gamma(L)$, *the language generated by* $\Gamma = (N, T, S, P)$. *There exists a leftmost derivation of* w.

Proof The proof uses induction of the number of steps n in the derivation. If $n = 1$, the derivation $S \Rightarrow w$ is obviously a leftmost derivation. If $n = k > 1$, and $S \Rightarrow^* w$, let $S \Rightarrow UV$ where $U, V \in (N \cup T)^*$. By Lemma 4.2, there exists derivations $U \Rightarrow^* w_1$ and $V \Rightarrow^* w_2$, where $w = w_1 w_2$. Since both of these derivations contain less than k steps, there exist leftmost derivations $U \Rightarrow^* w_1$ and $V \Rightarrow^* w_2$. Then $S \Rightarrow UV \Rightarrow^* w_1 V \Rightarrow^* w_1 w_2$ is a leftmost derivation of w. $\qquad\qquad \Box$

The following lemma shows that if the language L generated by a context-free grammar Γ does not contain the empty word λ then L can be generated by a context-free grammar which does not contain any productions of the form $A \rightarrow \lambda$, which we shall call a λ production. The only purpose of such a production is to remove A from the string of symbols. For example if we have productions $C \rightarrow abBaAaaa$ and $A \rightarrow \lambda$, we can then derive $C \Rightarrow^* abBaaaa$. We could simply remove $A \rightarrow \lambda$ and replace it with $abBaAaaa \rightarrow abBaaaa$. We would have to do this wherever A occurs in a production. For example if we have the $C \rightarrow abAaAba$, if $A \rightarrow \lambda$ is removed, we would have to include $C \rightarrow abAaba$, $C \rightarrow abaAba$, and $C \rightarrow ababa$.

Suppose we have productions $A \rightarrow aB$, $B \rightarrow C$, $C \rightarrow aa$, $C \rightarrow \lambda$. If we add the production $B \rightarrow \lambda$, we have created a new λ production. If we just remove $C \rightarrow \lambda$, we can no longer derive a. A nonterminal X is called **nilpotent** if $X \Rightarrow^* \lambda$. We solve the problem above by removing all nilpotents and not just those directly from λ productions. Thus we would also add the production $A \rightarrow a$ above since B is nilpotent.

In the next lemma it is necessary to be able to determine the nilpotents of a grammar. The following algorithm determines the set Θ of all nilpotents in a grammar by examining the productions.

(1) If $A \rightarrow \lambda$ is a production, then $A \in \Theta$.
(2) If $A \rightarrow A_1 A_2 \ldots A_n$ where $A_1, A_2, \ldots, A_n \in \Theta$, then $A \in \Theta$.
(3) Continue (2) until no new elements are added to Θ.

Lemma 4.4 *Let Γ be a grammar such that $\Gamma(L)$ does not contain the empty word. Form a grammar Γ' by beginning with the productions in Γ and*

(i) removing all λ productions;

(ii) for each production $A \to w$ in Γ, where $w = w_1 X_1 w_2 X_2 \ldots w_n X_n$ and (not necessarily distinct) nilpotents $X_1, X_2, \ldots, X_n$ in w, let P be the power set of $\{1, 2, \ldots, n\}$ and for $p \in P$, form productions $A \to w_p$ where w_p is the string w with the $\{X_i : i \in p\}$ removed, which produce λ productions.

The language generated by Γ' is equal to the language generated by Γ.

Proof Let L be the language generated by $\Gamma = (N, T, S, P)$, and L' be the language generated by $\Gamma' = (N, T, S, P')$. The language $L' \subseteq L$, since any production in Γ' which is not in Γ can be replaced with the original productions in Γ used to define it.

To show $L \subseteq L'$, let $w \in L$. Using induction on the number of steps in the derivation, we show that if $S \Rightarrow^* w$ using productions in P, then $S \Rightarrow^* w$ using productions in P'. If $n = 1$ then $S \Rightarrow w$ is obviously a production in P' since $w \neq \lambda$. Assume $n = k$ and $S \Rightarrow^* w$ is a derivation in Γ containing k steps. Let $S \Rightarrow A_1 A_2 A_3 \ldots A_m$ be the first derivation where $A_i \in N \cup T$. Therefore

$$S \Rightarrow A_1 A_2 A_3 \ldots A_m \Rightarrow^* w \text{ is the derivation } S \Rightarrow^* w.$$

By Lemma 4.2, there exist derivations $A_i \Rightarrow^* w_i$ in Γ, for $1 \leq i \leq m$, where $w = w_1 w_2 \ldots w_m$ and each derivation has less than k steps. By induction, if $w_i \neq \lambda$ there exists derivations $A_i \Rightarrow^* w_i$ in Γ' for $1 \leq i \leq m$ (note that if A_i is a terminal then $A_i = w_i$ and $A_i \Rightarrow^* w_i$ has 0 steps). Let $A_i' = A_i$ if $w_i \neq \lambda$ and $A_i' = \lambda$ if $w_i = \lambda$. Then $S \Rightarrow A_1' A_2' A_3' \ldots A_m'$ is a derivation in Γ' and

$$S \Rightarrow A_1' A_2' A_3' \ldots A_m' \Rightarrow^* w_1 A_2' A_3' \ldots A_m' \Rightarrow^* w_1 w_2 A_3' \ldots A_m' \Rightarrow^* w_1 w_2 \ldots w_m$$

is a derivation in Γ'. $\square$

For future reference, we point out that if one follows the proof above carefully, one finds that if $\Gamma(L)$ had contained the empty word, the only difference between $\Gamma(L)$ and $\Gamma(L')$ is that $\Gamma(L)$ would have contained the empty word, while $\Gamma(L')$ does not. The nonterminal S is a nilpotent that does not get removed, however if $S \to \lambda$ is a production, the production is removed.

We now start making progress toward Chomsky normal form. We show that given a grammar we can determine another grammar which generates the same language and has no productions of the form $A \to B$, where $A, B \in N$. These productions are called **trivial productions** since they simply relabel nonterminals. The process is simple. If $A \to B$ and $B \to W$, where $W \in (T \cup$

$N)^*$, we remove $A \rightarrow B$ and include $A \rightarrow W$. More generally, if $A_1 \rightarrow A_2 \rightarrow A_3 \rightarrow \cdots \rightarrow A_m \Rightarrow^* B$, where each $A_i \rightarrow A_{i+1}$ is a trivial production and $B \rightarrow w$, then remove the trivial productions and include $A_1 \rightarrow w$.

Lemma 4.5 *If $\Gamma(L)$, the language generated by $\Gamma = (N, T, S, P)$, does not contain the empty word, then there exists a grammar Γ' with no λ productions and no trivial productions such that $\Gamma(L) = \Gamma(L')$.*

Proof First assume Γ has had the λ productions removed as shown in the previous theorem. Create Γ' by removing all of the trivial projections and wherever $A_1 \rightarrow A_2 \rightarrow A_3 \rightarrow \cdots \rightarrow A_m \Rightarrow^* B$ occurs, where each $A_i \rightarrow A_{i+1}$ is a trivial production and $B \rightarrow w$, then remove the trivial productions and include $A_1 \rightarrow B$.

By construction, $\Gamma(L') \subseteq \Gamma(L)$.

Conversely, assume $S \Rightarrow^* w$ occurs in Γ. We use induction on the number of trivial derivations to show that there is a derivation $S \Rightarrow^* w$ in Γ'. Obviously if there is no trivial production then the derivation is in Γ'. Assume there are k trivial productions in the derivation. Assume that the derivation is a leftmost derivation of w. Assume $S \Rightarrow^* w$ has the form

$$
\begin{aligned}
S &\Rightarrow^* V_1 A_1 V_2 \rightarrow w_1 A_1 V_2 \rightarrow w_1 A_2 V_2 \rightarrow w_1 A_3 V_2 \rightarrow \cdots \rightarrow w_1 A_m V_2 \\
&\Rightarrow^* w_1 w' V_2 \\
&\Rightarrow w_1 w' w_2
\end{aligned}
$$

where $A_1 \rightarrow A_2 \rightarrow A_3 \rightarrow \cdots \rightarrow A_m$ is the last sequence of trivial productions in the derivation, and $A_m \rightarrow w'$. Then there are derivations $V_1 \Rightarrow^* w_1$, $V_2 \Rightarrow^* w_2$ in Γ, and

$$
S \Rightarrow^* V_1 A_1 V_2 \Rightarrow^* w_1 A_1 V_2 \Rightarrow w_1 w' V_2 \Rightarrow^* w_1 w' w_2
$$

has less trivial productions and all productions are in $\Gamma \cup \Gamma'$. Hence by the induction hypothesis there is a derivation $S \Rightarrow^* w$ in Γ'. □

Lemma 4.6 *If $\Gamma(L)$, the language generated by $\Gamma = (N, T, S, P)$, does not contain the empty word, then there exists a grammar $\Gamma' = (N, T, S, P')$ in which every production either has the form $A \rightarrow A_1 A_2 A_3 \ldots A_m$ for $n \geq 2$ where $A, A_1, A_2, A_3, \ldots, A_m$ are nonterminals or $A \rightarrow a$ where A is a nonterminal and a is a terminal such that $\Gamma(L) = \Gamma(L')$.*

Proof Assume all λ productions and all trivial productions have been eliminated. Thus all productions are of the form $A \rightarrow A_1, A_2, A_3, \ldots, A_m$ where $m \geq 2$ and $A_i \in N \cup T$ or $A \rightarrow a$ where A is a nonterminal and a is a terminal. If $A_1, A_2, A_3, \ldots, A_m$ all are nonterminals then the production has the proper form. If not, for each $A_i = a_i$ where a_i is a terminal, form a new nonterminal X_{a_i}. Replace $A \to A_1, A_2, A_3, \ldots, A_m$ with $A'_1, A'_2, A'_3, \ldots, A'_m$ where $A'_i = A_i$ if A_i is a nonterminal and $A'_i = X_{a_i}$ if A_i is a terminal. Thus if we have $V_1 a_1 V_2 a_2 V_3 a_3 \ldots V_n a_n V_{n+1}$ where $V_i \in N^*$ and a_i is a terminal, replace it with $V_1 X_{a_1} V_2 X_{a_2} V_3 X_{a_3} \ldots V_n X_{a_n} V_{n+1}$ and add productions $X_{a_i} \to a_i$ for $1 \le i \le n$. Let $\Gamma' = (N, T, S, P')$ be the new grammar formed. We need to show that $\Gamma(L') = \Gamma(L)$. Clearly $\Gamma(L) \subseteq \Gamma(L')$ since

$$A \Rightarrow V_1 a_1 V_2 a_2 V_3 a_3 \ldots V_n a_n V'_{n+1}$$

in Γ can be replaced by

$$\begin{aligned}
A &\Rightarrow V_1 X_{a_1} V_2 X_{a_2} V_3 X_{a_3} \ldots V_n X_{a_n} V_{n+1} \\
&\Rightarrow V_1 a_1 V_2 X_{a_2} V_3 X_{a_3} \ldots V_n X_{a_n} V_{n+1} \\
&\Rightarrow V_1 a_1 V_2 a_2 V_3 X_{a_3} \ldots V_n X_{a_n} V_{n+1} \\
&\vdots \\
&\Rightarrow V_1 a_1 V_2 a_2 V_3 a_3 \ldots V_n a_n V_{n+1}
\end{aligned}$$

in Γ'.

Conversely assume that in the derivation

$$S \Rightarrow^* w_1 w w_2$$

where $U \Rightarrow^* w_1$ and $A \Rightarrow^* w$ and $V_i \Rightarrow v_i$ are productions in Γ

$$\begin{aligned}
S &\Rightarrow^* UAU' \Rightarrow UV_1 X_{a_1} V_2 X_{a_2} \ldots V_n X_{a_n} V_{n+1} V \\
&\Rightarrow^* w_1 v_1 a_1 v_2 a_2 v_3 a_3 \ldots v_n a_n v_{n+1} w_2 = w_1 w w_2,
\end{aligned}$$

where

$$A \to V_1 X_{a_1} V_2 X_{a_2} \ldots V_n X_{a_n} V_{n+1}$$

is a production which is in Γ' and not in Γ and the derivation is

$$\begin{aligned}
S &\Rightarrow^* UAV \\
&\Rightarrow^* w_1 AV \\
&\Rightarrow w_1 V_1 X_{a_1} V_2 X_{a_2} \ldots V_n X_{a_n} V_{n+1} V \\
&\Rightarrow^* w_1 v_1 X_{a_1} V_2 X_{a_2} \ldots V_n X_{a_n} V_{n+1} V \\
&\Rightarrow w_1 v_1 a_1 V_2 X_{a_2} \ldots V_n X_{a_n} V_{n+1} V \\
&\Rightarrow^* w_1 v_1 a_1 v_2 X_{a_2} \ldots V_n X_{a_n} V_{n+1} V \\
&\Rightarrow w_1 v_1 a_1 v_2 a_2 V_3 X_{a_3} \ldots V_n X_{a_n} V_{n+1} V \\
&\vdots \\
&\Rightarrow w_1 v_1 a_1 v_2 a_2 v_3 a_3 \ldots v_n a_n V_{n+1} V \\
&\Rightarrow^* w_1 v_1 a_1 v_2 a_2 v_3 a_3 \ldots v_n a_n v_{n+1} V \\
&\Rightarrow^* w_1 w w_2.
\end{aligned}$$

This may be replaced by

$$S \Rightarrow^* U V_1 a_1 V_2 a_2 \ldots V_n a_n V_{n+1} V$$
$$\Rightarrow^* w_1 V_1 a_1 V_2 a_2 \ldots V_n a_n V_{n+1} V$$
$$\Rightarrow^* w_1 v_1 a_1 V_2 a_2 \ldots V_n a_n V_{n+1} V$$
$$\vdots \quad \vdots$$
$$\Rightarrow^* w_1 v_1 a_1 v_2 a_2 \ldots a_n V_{n+1} V$$
$$\Rightarrow^* w_1 v_1 a_1 v_2 a_2 \ldots a_n v_{n+1} V$$
$$\Rightarrow^* w_1 v_1 a_1 v_2 a_2 \ldots a_n v_{n+1} w_2$$
$$\Rightarrow^* w_1 w w_2.$$

We have a derivation for $S \Rightarrow^* w_1 w w_2$ in Γ. Hence $\Gamma' \subseteq \Gamma$. $\qquad \square$

From the above lemmas we are now able to prove that a context-free grammar Γ whose language does not contain the empty word can be expressed in **Chomsky normal form**.

Lemma 4.7 *If* $\Gamma(L)$, *the language generated by* $\Gamma = (N, T, S, P)$, *does not contain the empty word, then there exists a grammar in which every production has either the form*

$$A \to BC$$

or

$$A \to a$$

where A, B, and C are nonterminals and a is a terminal such that $\Gamma(L) = \Gamma(L')$.

Proof By the previous lemma, in which every production has either the form $A \to A_1 A_2 A_3 \ldots A_m$ where $A, A_1, A_2, A_3, \ldots, A_m$ are nonterminals or $A \to a$ where A is a nonterminal and a is a terminal. We construct a new grammar by replacing every production of the form $A \to A_1 A_2 A_3 \ldots A_m$ by the set of productions $A \to A_1 X_1, X_1 \to A_2 X_2, \ldots, X_{m-2} \to A_{m-1} A_m$, where each replacement of a production in Γ uses a new set of symbols.

$$A \Rightarrow A_1 X_2 \Rightarrow A_1 A_2 X_3 \Rightarrow^* A_1 A_2 A_3 \ldots A_m$$

is a derivation in Γ', $\Gamma(L) \subseteq \Gamma(L')$.

Conversely, if $S \Rightarrow^* w$ in Γ' contains no productions which are not in Γ, then $w \in \Gamma(L)$. If it does, let W_m be the last term in the derivation containing a symbol in Γ' which is not in Γ so we have $W_m \Rightarrow W_{m+1} \Rightarrow^* w$ and $W_m \Rightarrow W_{m+1}$ has the form $U' X_{m-2} V \Rightarrow U A_{m-1} A_m V$. Therefore the derivation uses the set of productions $A \to A_1X_1$, $X_1 \to A_2X_2$, ..., $X_{m-2} \to A_{m-1}A_m$ and has the form

$$
\begin{aligned}
S \Rightarrow^* UAV^* &\Rightarrow^* UA_1X_1V && \Rightarrow^* UA_1'X_1V \\
&\Rightarrow^* UA_1'A_2X_2V && \Rightarrow^* UA_1'A_2'X_2V \\
&\Rightarrow^* UA_1'A_2'A_3X_3V && \Rightarrow^* UA_1'A_2'A_3'X_3V \\
&\;\;\vdots \;\;\;\vdots && \;\;\vdots \;\;\;\vdots \\
&\Rightarrow^* UA_1' \cdots A_{m-2}X_{m-2}V && \Rightarrow^* UA_1' \cdots A_{m-2}'X_{m-2}V \\
&\Rightarrow \;\; W_{m+1} && \Rightarrow^* w
\end{aligned}
$$

where $U' = UA_1'A_2'A_3' \cdots A_{m-2}'$ and $A_i \Rightarrow^* A_i$ is a derivation in Γ. If this derivation $S \Rightarrow^* w$ is not in Γ, we again pick the last term in the derivation containing a symbol in Γ' which is not in Γ, and continue the process until no such terms are left. Therefore $w \in \Gamma$ and $\Gamma(L') \subseteq \Gamma(L)$. $\qquad\square$

Finally we remove the restriction that $\Gamma(L)$ contains the empty word. As mentioned, following the proof of Lemma 4.4, by eliminating λ productions, if $\Gamma(L)$ contained the empty word, one produced the same language with only the empty word eliminated. Since all of the languages of the forms of grammars developed since Lemma 4.4 are the same, if $\Gamma(L)$ contained the empty word, the language developed by the grammar Γ' in the previous lemma would have differed from $\Gamma(L)$ only in the fact that $\Gamma(L)$ contained the empty word while $\Gamma(L')$ did not. Thus to get $\Gamma(L)$ we need only have productions that add the empty word to the language and leave the rest of the language alone. We do this by adding two new nonterminal symbols, S' and ψ, where S' is the new start symbol and productions $S' \to S\psi$ and $\psi \to \lambda$. Call this the λ **extended Chomsky normal form**.

Theorem 4.4 *Given a context-free grammar Γ containing the empty word, there is a context-free grammar Γ' in λ extended Chomsky normal form so that $\Gamma(L) = \Gamma(L')$.*

We now consider converting a context-free language to Greibach normal form. Even though we use leftmost derivations, we have no bound on how many derivations may occur before the first terminal symbol appears at the left of the string. For example, using the production $A \to Aa$, we can generate the string Aa^n for arbitrary n, using n derivations without beginning a string with a terminal symbol. We can eliminate this particular problem by eliminating the productions of the form $A \to Aa$. This is called **elimination of left recursion**. In a grammar Γ with no λ productions or trivial productions, let

$$A \to AV_1, A \to AV_2, \ldots, A \to AV_n$$

be productions in which the right-hand side of the production begins with an A and

$$A \to U_1, A \to U_2, \ldots, A \to U_m$$

be productions in which the right-hand side of the production does not begin with an A. We form a new grammar Γ' by adding a new nonterminal A' to the grammar: and using the following steps:

(1) Eliminate all productions of the form $A \to AV_i$ for $1 \le i \le n$.
(2) Form productions $A \to U_i A'$ for $1 \le i \le n$.
(3) Form productions $A' \to V_i A'$ and $A' \to V_i$.

Lemma 4.8 $\Gamma(L) = \Gamma'(L)$.

Proof Let a derivation beginning with A have the form assuming $A \to U_i A'$ for $1 \le i \le n$ and we have $A \Rightarrow AV_{(1)} \Rightarrow AV_{(2)}V_{(1)} \Rightarrow^* AV_{(k)} \ldots V_{(2)}V_{(1)} \Rightarrow U_{(i)}V_{(k)} \ldots V_{(2)}V_{(1)}$ where $V_{(j)} \in \{V_1, V_2, \ldots, V_n\}$ for all $1 \le j \le k$ and $U_{(i)} \in \{U_1, U_2, \ldots, U_m\}$. Therefore using leftmost derivation, any production containing A will have the form

$$wAW \Rightarrow wAV_{(1)}W \Rightarrow wAV_{(2)}V_{(1)}W \Rightarrow^* wAV_{(k)} \ldots V_{(2)}V_{(1)}W$$
$$\Rightarrow wU_{(i)}V_{(k)} \ldots V_{(2)}V_{(1)}W.$$

But

$$A \Rightarrow AV_{(1)} \Rightarrow AV_{(2)}V_{(1)} \Rightarrow^* AV_{(k)} \ldots V_{(2)}V_{(1)} \Rightarrow U_{(i)}V_{(k)} \ldots V_{(2)}V_{(1)}$$

can be replaced by

$$A \Rightarrow U_{(i)}A' \Rightarrow U_{(i)}V_{(k)}A' \Rightarrow U_{(i)}V_{(k)}V_{(k-1)}A'$$
$$\Rightarrow^* U_{(i)}V_{(k)} \ldots V_{(2)}A' \Rightarrow U_{(i)}V_{(k)} \ldots V_{(2)}V_{(1)}.$$

Placing w on the left and W on the right of each term, we have, $wAW \Rightarrow^* wU_{(i)}V_{(k)} \ldots V_{(2)}V_{(1)}W$ in Γ'. Hence $\Gamma(L) \subseteq \Gamma'(L)$.

The proof that $\Gamma'(L) \subseteq \Gamma(L)$ is left to the reader. □

Lemma 4.9 *Let $A \to UBV$ be a production in Γ and $B \to W_1, B \to W_2, \ldots, B \to W_m$ be the set of all productions in Γ with B on the left. Let Γ' be the grammar with production $A \to UBV$ removed and the productions $A \to UW_i V$ for $1 \le i \le m$ added, then $\Gamma(L) = \Gamma'(L)$.*

Proof The production $A \to UW_i V$ can always be replaced by the production $A \to UBV$ followed by the production $B \to W_i$. Hence $\Gamma'(L) \subseteq \Gamma(L)$. The proof that $\Gamma(L) \subseteq \Gamma'(L)$ is left to the reader. □

The theorem that every context–free grammar can be expressed in Greibach normal form can be proved by first expressing the grammar in Chomsky normal form. We shall not do so however so that the development for Chomsky normal form may be omitted if desired. Using the above lemmas, we are about to take a giant leap toward proving that every context-free grammar can be expressed in Greibach normal form.

Lemma 4.10 *Any context-free grammar which does not generate* λ *can be expressed so that each of its productions is of the form*

$$A \to aW,$$

where a is a terminal and W is a string which is empty or consists of a string of terminals and/or nonterminals.

Proof We first order the nonterminals beginning with S, the start symbol. For simplicity, let the nonterminals be $A_1, A_2, A_3, \ldots, A_m$. Our first goal is to change every production so that it is either in the form

$$A \to aW,$$

where a is a terminal and W is a string which is empty or consists of a string of terminals and/or nonterminals, or in the form

$$A_i \to A_j Y,$$

where $i < j$ and Y consists of a string of terminals and/or nonterminals. *Recall that using the procedures for elimination of left recursion and for eliminating a nonterminal described in Lemma 4.9 to alter the productions of the grammar does not change the language generated by the grammar.*

Using induction, for $i = 1$, since $S = A_1$ is automatically less than every other nonterminal, we need only consider $S \to SY$, the S on the right-hand side of the production can be removed by the process of elimination of left recursion. Assume it is true for every A_i where $i < k$. We now prove the statement for $i = k$. In each case where $A_k \to A_j Y$ is a production for $k > j$, use the procedure in Lemma 4.9 to eliminate A_j. When $A_k \to A_k Y$ is a production, use the process of elimination of left recursion to remove A_k from the right-hand side.

Therefore by induction we have every production so that it is either in the form

$$A \to aW,$$

where a is a terminal and W is a string which is empty or consists of a string of terminals and/or nonterminals, or in the form

$$A_i \rightarrow A_j Y,$$

where $i < j$ and Y consists of a string of terminals and/or nonterminals.

Any production with A_m on the left-hand side must have the form $A_m \rightarrow aW$ since there is no nonterminal larger than A_m. If there is a production of the form $A_{m-1} \rightarrow A_m W'$, use the procedures in Lemma 4.9 to eliminate A_m. The result is a production of the form $A_m \rightarrow bW''$. Assume k is the largest value so $A_k \rightarrow A_j Y$ is a production where $k < j$. Again using the procedures in Lemma 4.9 to eliminate A_j, we have a procedure of the form $A_k \rightarrow aW$. When the process is completed, we have

$$A_i \rightarrow aW$$

where a is a terminal and W is a string which is empty or consists of a string of terminals and/or nonterminals for every i. We now have to consider the B_i created using a process of elimination of left recursion. From the construction of the B_i, it is impossible to have a production of the form $B_i \rightarrow B_j W$. Therefore productions with B_i on the left have the form $B_i \rightarrow aW$ or $B_i \rightarrow A_j W$. Repeating the process above we can change these to the form $B_i \rightarrow aW$, and the lemma is proved. $\qquad\square$

Theorem 4.5 *Every context-free grammar whose language does not contain λ can be expressed in Greibach normal form.*

Proof We outline the proof. The details are left to the reader. Since we already know that every production can be written in the form

$$A \rightarrow aW,$$

where a is a terminal and W is a string which is empty or consists of a string of terminals and/or nonterminals, for every terminal b in W replace it with nonterminal A_b and add the production $A_b \rightarrow b$. Hint: see proof of Lemma 4.6. $\qquad\square$

For any context free grammarcontaining the empty word we can form a grammar in **extended Greibach normal form**, which accepts the empty word by simply adding the production $S \rightarrow \lambda$ after the completion of the Greibach normal form.

Exercises

(1) In Lemma 4.9 "Let $A \to UBV$ be a production in Γ and $B \to W_1, B \to W_2, \ldots, B \to W_m$ be the set of all productions in Γ with B on the left. Let Γ' be the grammar with production $A \to UBV$ removed and the productions $A \to UW_iV$ for $1 \le i \le m$ added, then $\Gamma(L) = \Gamma'(L)$," prove $\Gamma(L) \subseteq \Gamma'(L)$.

(2) Prove Theorem 4.5 "Every context-free grammar can be expressed in Greibach normal form."

(3) Complete the proof of Lemma 4.8.

(4) Let $\Gamma'' = (N, \Sigma, S, P)$ be the grammar defined by $N = \{S, A, B\}$, $\Sigma = \{a, b\}$, and P be the set of productions

$$S \to ABABABA \qquad A \to Aa \qquad A \to \lambda \qquad B \to b.$$

Express this grammar in Chomsky normal form.

(5) Express the previous grammar in Greibach normal form.

(6) Let $\Gamma = (N, T, S, P)$ be the grammar with $N = \{S\}$, $T = \{a, b\}$, and P contain the productions

$$S \to SS \qquad B \to aa \qquad S \to BS \qquad B \to bb \qquad S \to SB$$
$$A \to ab \qquad S \to \lambda \qquad A \to ba \qquad S \to ASA.$$

Express this grammar in Chomsky normal form.

(7) Express the previous grammar in Greibach normal form.

(8) Let $\Gamma'' = (N, \Sigma, S, P)$ be the grammar defined by $N = \{S, A, B\}$, $\Sigma = \{a, b\}$, and P be the set of productions

$$S \to AbaB \qquad A \to bAa \qquad A \to \lambda \qquad B \to AAb \qquad BaabA.$$

Express this grammar in Chomsky normal form.

(9) Express the previous grammar in Greibach normal form.

4.3 Pushdown automata and context-free languages

The primary importance of the PDA is that a language is accepted by a PDA if and only if it is constructed by context-free grammar. Recall that a context-free language is a language that is generated by a context-free grammar. In the remainder of this section, we show that a language is context-free if and only if it is accepted by a PDA.

We first demonstrate how to construct a PDA that will read the language generated by a context-free grammar.

Before beginning we need two tools. The first is the concept of pushing a string of stack symbols into the stack. We are not changing the definition of the stack. To push a string into the stack we simply mean that we are pushing the last symbol of the string into the stack, then the next to last symbol into the stack, and continuing until the first symbol of the string has been pushed into the stack. If the string is then popped a symbol at a time, the symbols form the original string. For example to push $abAc$, we first push c, then push A, then push b, and finally push a. Thus we may consider a PDA to have the form $M = \{\Sigma, Q, s, I, \Upsilon, F)$ where Σ is the alphabet, Q is the set of states, s is the initial or starting state, I is the set of stack symbols, F is the set of acceptance states, and Υ is the transition relation. The relation Υ is a finite subset of $((Q \times \Sigma^* \times I^*) \times (Q \times I^*))$. This means that the machine in some state $q \in Q$ can read a possibly empty string of the alphabet by reading a letter at a time if the string is nonempty, pop and read a possibly empty string of symbols by popping and reading a symbol at a time if the string of symbols in nonempty and as a result can read a string of letters, change state, and push a string of symbols onto the stack as described above.

Throughout the remainder of this section we shall assume that only left derivations for context-free languages are used and that the PDA has only two states, s and t. The alphabet Σ in the PDA consist of the terminal symbols of the grammar Γ. The stack symbols of the PDA consist of the terminal and nonterminal symbols of the grammar Γ, i.e. $I = T \cup N$.

To convert a context-free grammar Γ into a PDA, which accepts the same language generated by Γ, we use the following rules:

(1) Begin by pushing S, start symbol of the grammar, i.e. begin with the automaton.

(2) If a nonterminal A is popped from the stack, then for some production $A \rightarrow w$ in Γ, w is pushed into the stack, i.e. we have the automaton in

figure

(3) If a terminal *a* is popped from the stack then *a* must be read, i.e. if we have the automaton

then we have the automaton

Thus the terminal elements at the top of the stack are removed and matched with the letters from the tape.

The result is that we are imitating the productions in the grammar by popping the first part of the production and pushing the second part so that while the grammar replaces the first part of the production with the second part, so does the PDA. The stack then resembles the strings derived in the grammar except that the terminals on the left of the derived string (top of the stack) are then removed as they occur in the stack and compared with the letters on the tape. As before a word is accepted if the word has been read and the stack is empty.

Example 4.17 Let $\Gamma = (N, T, S, P)$ be the grammar with $N = \{S\}$, $T = \{a, b\}$, and P contain the productions

$$S \to aSa \qquad S \to bSb \qquad S \to \lambda$$

which generates the language $\{ww^R : w \in T^*\}$. This has the PDA

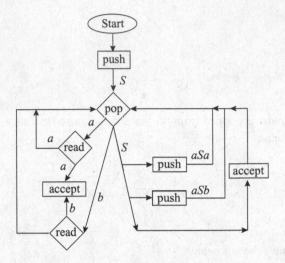

Consider the word *abba*. We can trace its path with the following table:

state	instruction	stack	tape	state	instruction	stack	tape
x_0	start	λ	abba	t	pop b	Sba	bba
t	push S	S	abba	t	read b	Sba	ba
t	pop S	λ	abba	t	pop S	ba	ba
t	push aSa	aSa	abba	t	pop b	a	ba
t	pop a	Sa	abba	t	read b	a	a
t	read a	Sa	bba	t	pop a	a	λ
t	pop S	a	bba	t	read a	λ	λ
t	push bSb	$bSba$	bba	t	accept	λ	λ

Example 4.18 Let $\Gamma = (N, T, S, P)$ be the grammar with $N = \{S\}$, $T = \{a, b\}$, and P contain the productions

$$S \to SS \qquad B \to aa \qquad S \to BS \qquad B \to bb \qquad S \to SB$$
$$A \to ab \qquad S \to \lambda \qquad A \to ba \qquad S \to ASA$$

which generates the language $\{w : w \in A^*$ and contains an even number of as and an even number of bs.$\}$. This has the PDA

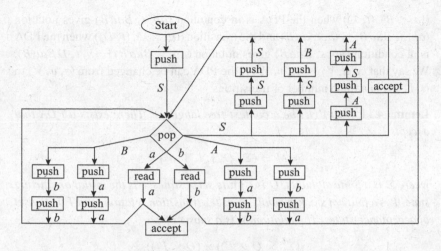

Consider the word *abbabb*. We can trace its path with the following table. In this table to save space, strings will be pushed in as one operation rather than pushing in each symbol.

instruction	stack	tape	instruction	stack	tape
start	λ	abbabb	pop	AS	babb
push S	S	abbabb	pop	S	babb
pop	λ	abbabb	push ba	baS	babb
push SS	SS	abbabb	pop	aS	babb
pop	S	abbabb	read a	S	bb
push ASA	ASAS	abbabb	pop	λ	bb
pop	SAS	abbabb	push bb	bb	bb
push ab	abSAS	abbabb	pop	b	bb
pop	bSAS	abbabb	read b	b	b
read a	bSAS	bbabb	pop	λ	b
pop	SAS	bbabb	read b	λ	λ
read b	SAS	babb			

Before formally proving that a language $\Gamma(L)$ is context-free if and only if it is accepted by a PDA, we shall adopt a notation for PDAs which will be more convenient. We shall denote by an ordered triple the current condition of the PDA. This triple consists of the current state of the machine, the remaining string of input symbols to be read, and the current string in the stack. For example the triple $(s, aabb, AaBaB)$ represents the PDA in state s, with $aabb$ on the input tape, and $AaBaB$ in the stack. Given triples (s, u, V) and (t, v, W), then notation $(s, u, V) \vdash (t, v, W)$ indicates that the PDA can be changed from (s, u, V) to (t, v, W) in a single transition from F. For example the transition

$((a, s, B), (t, \lambda))$ when the PDA is in condition $(ab, s, Baa B)$ gives notation $(ab, s, Baa B) \vdash (b, t, aa B)$ and the transition $((a, s, \lambda), (t, D))$ when the PDA is in condition $(ab, s, Baa B)$ gives notation $(ab, s, Baa B) \vdash (b, t, DBaa B)$. We say that $(s, u, V) \vdash^* (t, v, W)$ if the PDA can be changed from (s, u, V) to (t, v, W) in a finite number of transitions.

Lemma 4.11 *Let $\Gamma(L)$ be a context-free language. There exists a PDA that accepts L*

$$M = (\Sigma, Q, s, I, \Upsilon, F)$$

where Σ is a finite alphabet, Q is a finite set of states, s is the initial or starting state, I is a finite of stack symbols, Υ is the transition relation, and F is the set of acceptance states. The relation Υ is a subset of

$$((\Sigma^\lambda \times Q \times I^\lambda) \times (Q \times I^\lambda)).$$

Proof As previously mentioned, we shall assume that the PDA has two states which we shall denote here as s and t so that $M = \{\Sigma, Q, s, I, \Upsilon, F\}$ where Σ, the alphabet, consists of the terminal symbols T of the grammar $\Gamma = (N, T, S, P)$, $Q = \{s, t\}$, s is the initial or starting state, I consists of the terminal and nonterminal symbols of the grammar, i.e. $I = T \cup N$, the set of stack symbols, $T = \{s, t\}$, and Υ is the transition relation defined as follows:

(1) $((s, \lambda, \lambda), (t, S)) \in \Upsilon$ so $(s, u, \lambda) \vdash (t, u, S)$ for $u \in T^*$. Begin by pushing S, start symbol of the grammar, i.e. The automaton begins with

(2) If $A \to w$ in Γ, then $(t, \lambda, A), (t, w)) \in \Upsilon$ so $(t, u, A) \vdash (t, u, w)$ for $u \in T^*$. (If a nonterminal A is popped from the stack, then for some production $A \to w$ in Γ, w is pushed into the stack, i.e. we have in the automaton

(3) For all $a \in T$, $((t, a, a), (t, \lambda)) \in \Upsilon$ so $(t, au, aw) \vdash (t, u, w)$ for $u \in T^*$. If a terminal a is popped from the stack then a must be read, i.e. if we have in the automaton

it must be followed by

We shall assume that Γ is in Chomsky normal form. The reader is asked to prove the theorem when Γ is in Greibach normal form.

Using left derivation, every string that is derived either begins with a terminal or contains no terminal. In the corresponding PDA, the derived string is placed in the stack using (2). If there is a terminal, it is compared with the next letter to be read as input. If they agree, then the terminal is removed from the stack using (3). If the word generated by the terminal is the same as the word generated by grammar, each terminal will be removed from the stack as it is generated leaving only an empty stack after the tape has been read.

We first show, assuming leftmost derivation in Γ, that if $S \Rightarrow^* \alpha\beta$ where $\alpha \in T^*$ and β begins with a nonterminal or is empty, then $(t, \alpha, S) \vdash^* (t, \lambda, \beta)$. Hence if $S \Rightarrow^* \alpha$ in Γ where $\alpha \in T^*$, then $(s, \alpha, \lambda) \vdash (t, \alpha, S) \vdash^* (t, \lambda, \lambda)$, and α is accepted by the PDA. We prove this using induction on the length of the derivation. Suppose $n = 0$, but then we have $S \Rightarrow^* S$, so $\alpha = \lambda$, $\beta = S$, and $(t, \lambda, S) \vdash^* (t, \lambda, S)$ gives us $(t, \alpha, S) \vdash^* (t, \lambda, \beta)$. Now assume $S \Rightarrow^* \gamma$ in $k + 1$ steps. Say

$$S \Rightarrow m_1 \Rightarrow m_2 \Rightarrow^* m_k \Rightarrow m_{k+1}.$$

Then there is a first nonterminal B in the string m_k and a production $B \to w$ so $m_k = uBv$ and $m_{k+1} = uwv$. By the induction hypothesis, since $S \Rightarrow^* uBv$, $(t, u, S) \vdash^* (t, \lambda, Bv)$. Since $B \to w$ using relation (2), we have $(t, \lambda, B), (t, w)) \in \Upsilon$ and $(t, \lambda, Bv) \vdash (t, \lambda, wv)$. If the production $B \to w$ has the form $B \to CD$, where C and D are nonterminals, so that $w = CD$, then w begins with a nonterminal and $(t, u, S) \vdash^* (t, \lambda, Bv) \vdash (t, \lambda, wv)$ or $(t, u, S) \vdash^* (t, \lambda, wv)$ where $m_{k+1} = uwv$ as desired.

If the production has the form $B \to a$ so that $w = a$, where a is a terminal, then $(t, ua, S) \vdash^* (t', a, Bv) \vdash (t, a, av) \vdash (t, \lambda, v)$ using derivation (3) above. Hence $(t, uw, S) \vdash^* (t, \lambda, v)$ where $m_{k+1} = uwv$ as desired. Note that since we are using leftmost derivation, v must begin with a nonterminal.

We now show that if $(t, \alpha, S) \vdash^* (t, \lambda, \beta)$ with $\alpha \in T^*, \beta \in I^*$, then $S \Rightarrow^*$ $\alpha\beta$. Hence if $(s, \alpha, \lambda) \vdash (t, \alpha, S) \vdash^* (t, \lambda, \lambda)$ so α is accepted by the PDA, then $S \Rightarrow^* \alpha$ so α is generated by the grammar.

We again use induction on the length of the computation by the PDA. If $k = 0$, then $(t, \lambda, S) \vdash^* (t, \lambda, S)$ so $S \Rightarrow^* S$, which is certainly true. Assume $(t, \alpha, S) \vdash^* (t, \lambda, \beta)$ in $k + 1$ steps so that $(t, \alpha, S) \vdash^* (t, v, \gamma)$ in k steps and $(t, v, \gamma) \vdash (t, \lambda, \beta)$. If the transition relation for $(t, w, \gamma) \vdash (t, \lambda, \beta)$ is relation (2), then $\gamma = Bv, \beta = wv$ and $B \to w$. Since no input is read, we have $w = \lambda$. Therefore by induction, $S \Rightarrow^* \alpha\gamma = \alpha Bv$. Since $B \to w, \alpha Bv \Rightarrow \alpha wv = \alpha\beta$. Therefore $S \Rightarrow^* \alpha\beta$. If the transition relation for $(t, v, \gamma) \vdash (t, \lambda, \beta)$ is type (3), then $v = a$ and $\gamma = a\beta$ for some terminal a. But since a is the last input read from the string $\alpha, \alpha = au$ for $u \in T^*$. Hence $(t, u, S) \vdash^* (t, \lambda, \gamma)$ and by induction $S \Rightarrow^* u\gamma = ua\beta = \alpha\beta$. $\square$

For a given pushdown automaton $M = (\Sigma, Q, s, I, \Upsilon, F)$, we next wish to construct a context-free grammar $\Gamma = (N, T, S, P)$. The expression N shall be of the form $\langle p, B, q \rangle$, where p and q are states of the automaton and B is in the stack. Thus $\langle p, B, q \rangle$ represents the input u read in passing from state p to state q, where B is removed from the stack. In fact we shall have $\langle p, B, q \rangle \Rightarrow^* u$. The terminal $\langle p, \lambda, q \rangle$ represents the input read in passing from state p to state q and leaving the stack as it was in state p. The productions consist of the following four types.

(1) For each $q \in T$, the production $S \to \langle s, \lambda, q \rangle$.
(2) For each transition $((p, a, B), (q, D)) \in \Upsilon$, where $B, D \in I \cup \{\lambda\}$, the productions $\langle p, B, t \rangle \to a\langle q, D, t \rangle$ for all $t \in Q$.
(3) For each transition $((p, a, D), (q, B_1 B_2 \ldots B_n)) \in \Upsilon$, where $D \in I \cup \{\lambda\}, B_1, B_2, \ldots, B_n \in C$, the productions

$$\langle p, D, t \rangle \to a\langle q, B_1, q_1 \rangle\langle q_1, B_2, q_2 \rangle\langle q_2, B_3, q_3 \rangle \ldots \langle q_{n-1}, B_n, t \rangle$$

for all $q_1, q_2, \ldots, q_{n-1}, t \in Q$.
(4) For each $q \in Q$, the production $\langle q, \lambda, q \rangle \to \lambda$.

The first statement intuitively says that at the beginning we need to generate the entire word accepted by the PDA. The second statement intuitively says that the output generated by $\langle p, B, t \rangle$, which is the input to be read by the PDA in state p using stack B moving to state t, is equal on the right-hand side of the production to the input read in passing from state p with stack A to state q with stack D followed by the output generated by $\langle q, D, t \rangle$ which is the input read by the PDA in state q with stack D moving to state t.

The third statement intuitively says that the output generated by $\langle p, D, t \rangle$, which is the input to be read by the PDA in state p with stack D moving to state t, is equal on the right-hand side of the production to the input read in passing from state p with stack D to state q with stack $B_1 B_2 \ldots B_n$ followed by the output generated by $\langle q, B_1, q_1 \rangle$, the input read by passing from state q using stack B_1 to state $q_1 \ldots$ followed by the output generated by $\langle q_1, B_2, q_2 \rangle$, the input read by passing from state q_1 using stack B_2 to state $q_2 \ldots$ followed by the output generated by $\langle q_{n-1}, B_n, t \rangle$, the input read by passing from state q_{n-1} using stack B_n to state t.

The fourth statement intuitively says that to move from a state to itself requires no input. Note that the productions of type (4) are the only ones which pop nonterminals without replacing them with other nonterminals. Hence a word in the language of the grammar cannot be generated without these productions.

Lemma 4.12 *A language $M(L)$ accepted by a pushdown automaton $M = (\Sigma, Q, s, I, \Upsilon, F)$, is a context-free language.*

Proof Using the grammar $\Gamma = (N, \Sigma, S, P)$, where the nonterminals and productions are described above, we show that Γ generates the same language as accepted by M.

We first show that for $p, q \in Q, B \in I \cup \{\lambda\}$ and $w \in A^*$, that

$$\langle p, B, q \rangle \Rightarrow^* w \text{ if and only if } (p, w, B) \vdash^* (q, \lambda, \lambda).$$

Thus for $t \in Q$, $\langle s, \lambda, t \rangle \Rightarrow^* w$ if and only if $(s, w, \lambda) \vdash^* (t, \lambda, \lambda)$ so that a word is generated by Γ if and only if it is accepted by M.

First, using induction on the number of derivation steps, we show that if $\langle p, B, q \rangle \Rightarrow^* w$ then $(p, w, B) \vdash^* (q, \lambda, \lambda)$. Beginning with $n = 1$, the only possibility is that a nonterminal is popped, without replacement. This can only occur using productions of type (4), so we have $p = q$, $B = \lambda$, and $w = \lambda$. But this gives us $(p, \lambda, \lambda) \vdash^* (p, \lambda, \lambda)$ which is obvious. Assume $n = k > 1$, then the first production can only be of type (2) or type (3). If it is type (2), we have $\langle p, B, q \rangle \to a \langle r, D, q \rangle$ for $p, r \in Q$, where $((p, a, B), (r, D)) \in \Upsilon$. Hence letting $w = av, (p, w, B) \vdash (q, v, D)$ and by induction if $\langle r, D, q \rangle \Rightarrow^* v$ then $(q, v, D) \vdash^* (q, \lambda, \lambda)$. Therefore $(p, w, B) \vdash^* (q, \lambda, \lambda)$.

If the first production is of type (3), we have

$$\langle p, B, q \rangle \Rightarrow a \langle q_0, B_1, q_1 \rangle \langle q_1, B_2, q_2 \rangle \langle q_2, B_3, q_3 \rangle \ldots \langle q_{n-1}, B_n, q \rangle \Rightarrow^* w$$

and $((p, a, B), (q, B_1 B_2 \ldots B_n)) \in \Upsilon$. So if $w = av$, $(p, w, B) \vdash (q, v, B_1 B_2 \ldots B_n)$. For convenience of notation, let $q = q_n$. Let $\langle q_{i-1}, B_i, q_i \rangle \Rightarrow^* u_i$ so that $w = au_1 u_2 \ldots u_n$ and $v = u_1 u_2 \ldots u_n$. By induction, $(q_{i-1}, u_i, B_i) \vdash^* (q_i, \lambda, \lambda)$.

Therefore we have

$$
\begin{aligned}
(p, w, B) \ &\vdash (q_0, u_1 u_2 \ldots u_n, B_1 B_2 \ldots B_n) \\
&\vdash^* (q_1, u_2 \ldots u_n, B_2 \ldots B_n) \\
&\vdash^* (q_2, u_3 \ldots u_n, B_3 \ldots B_n) \\
&\ \vdots \quad \vdots \\
&\vdash^* (q_{n-1}, u_n, B_n) \\
&\vdash (q_n, \lambda, \lambda)
\end{aligned}
$$

so that $(p, w, B) \vdash^* (q, \lambda, \lambda)$.

We now show that if $(p, w, B) \vdash^* (q, \lambda, \lambda)$ then $\langle p, B, q \rangle \Rightarrow^* w$. We use induction on the number of steps in $(p, w, B) \vdash^* (q, \lambda, \lambda)$. If there are 0 steps, then $p = q$ and $w = B = \lambda$. This corresponds to $\langle p, \lambda, p \rangle \Rightarrow \lambda$ which is one of the productions. Therefore the statement is true for 0 steps.

Assume $(p, w, B) \vdash^* (q, \lambda, \lambda)$ in $k + 1$ steps. First assume that we have $w = av$ and

$$(p, w, B) \vdash (q, v, D) \vdash^* (q, \lambda, \lambda)$$

where $((p, a, B), (r, D)) \in \Upsilon$, and $B, D \in I \cup \{\lambda\}$, giving productions $\langle p, B, q \rangle \to a \langle r, D, q \rangle$. Since $(q, v, D) \vdash^* (q, \lambda, \lambda)$ by induction hypothesis, $\langle r, D, q \rangle \Rightarrow^* v$. Therefore $\langle p, B, q \rangle \Rightarrow a \langle r, D, q \rangle \Rightarrow^* av = w$ and we are finished.

Next assume $w = av$ and the first step is $(p, w, B) \vdash (q, v, B_1 B_2 \ldots B_n)$ so we have

$$(p, w, B) \vdash (q_0, v, B_1 B_2 \ldots B_n) \vdash^* (q, \lambda, \lambda)$$

and each B_i is eventually removed from the stack in order so that there are states $q_1, q_2, \ldots q_{n-1}, q_n$ where $q_n = q$ and $v = v_1 v_2 \ldots v_{n-1} v_n$ such that

$$
\begin{aligned}
(p, w, B) \ &\vdash (q_0, v_1 v_2 \ldots v_{n-1} v_n, B_1 B_2 \ldots B_n) \\
&\vdash^* (q_1, v_2 \ldots v_{n-1} v_n, B_2 \ldots B_n) \\
&\vdash^* (q_2, v_3, \ldots v_{n-1} v_n, B_3 \ldots B_n) \\
&\ \vdots \quad \vdots \\
&\vdash^* (q_{n-1}, v_n, B_n) \\
&\vdash^* (q_n, \lambda, \lambda).
\end{aligned}
$$

By the induction hypothesis, $\langle q_{i-1}, B_i, q_i \rangle \Rightarrow^* v_i$.
 But since the production is type (3),

$$\langle p, B, q \rangle \Rightarrow a \langle q_0, B_1, q_1 \rangle \langle q_1, B_2, q_2 \rangle \langle q_2, B_3, q_3 \rangle \ldots \langle q_{n-1}, B_n, q \rangle$$
$$\Rightarrow^* a v_1 \langle q_1, B_2, q_2 \rangle \langle q_2, B_3, q_3 \rangle \ldots \langle q_{n-1}, B_n, q \rangle$$
$$\Rightarrow^* a v_1 v_2 \langle q_2, B_3, q_3 \rangle \ldots \langle q_{n-1}, B_n, q \rangle$$
$$\vdots \quad \vdots$$
$$\Rightarrow^* a v_1 v_2 \ldots v_{n-1} \langle q_{n-1}, B_n, q \rangle$$
$$\Rightarrow^* a v_1 v_2 \ldots v_{n-1} v_n$$

so that $\langle p, B, q \rangle \Rightarrow^* w$. $\square$

Theorem 4.6 *A language is context-free if and only if it is accepted by a PDA.*

Exercises

(1) Construct a pushdown automaton which reads the same language as the grammar $\Gamma = (N, \Sigma, S, P)$ defined by $N = \{S, A, B\}$, $\Sigma = \{a, b, c\}$, and the set of productions P given by

$$S \rightarrow aA \qquad A \rightarrow aAB \qquad A \rightarrow a \qquad B \rightarrow b \qquad B \rightarrow \lambda.$$

(2) Construct a pushdown automaton which reads the same language as generated by the grammar $\Gamma = (N, \Sigma, S, P)$ defined by $N = \{S, A, B\}$, $\Sigma = \{a, b, c\}$, and the set of productions P given by

$$S \rightarrow AB \qquad A \rightarrow abaA \qquad A \rightarrow \lambda \qquad B \rightarrow Bcacc \qquad B \rightarrow \lambda.$$

(3) Construct a pushdown automaton which reads the same language as generated by the grammar $\Gamma = (N, \Sigma, S, P)$ defined by $N = \{S, A, B\}$, $\Sigma = \{a, b, c, \}$, and the set of productions P given by

$$S \rightarrow AcB \qquad A \rightarrow abaA \qquad A \rightarrow \lambda \qquad B \rightarrow Bcacb \qquad B \rightarrow \lambda.$$

(4) Construct a pushdown automaton which reads the same language as generated by the grammar $\Gamma = (N, \Sigma, S, P)$ defined by $N = \{S, A, B\}$, $\Sigma = \{a, b, c\}$, and the set of productions P given by

$$S \rightarrow AB \qquad A \rightarrow acA \qquad B \rightarrow bcB \qquad B \rightarrow bB$$
$$A \rightarrow aAa \qquad B \rightarrow \lambda \qquad A \rightarrow \lambda.$$

(5) Construct a pushdown automaton which reads the same language as generated by the grammar $\Gamma = (N, \Sigma, S, P)$ defined by $N = \{S, A, B\}$, $\Sigma = \{a, b, c, d\}$, and the set of productions P given by

$$S \to AB \qquad A \to aAc \qquad B \to bBc \qquad B \to bB$$
$$A \to AaA \qquad B \to \lambda \qquad A \to \lambda.$$

(6) Construct a grammar which generates the language read by the pushdown automaton

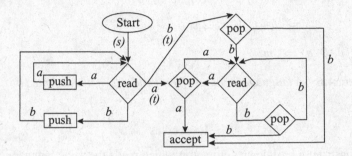

(7) Construct a grammar which generates the language read by the pushdown automaton

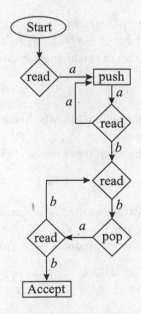

(8) Construct a grammar which generates the language read by the pushdown automaton

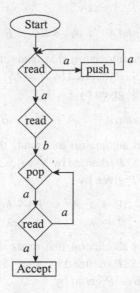

(9) Construct a grammar which generates the language read by the pushdown automaton

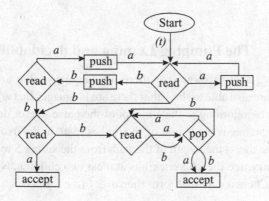

(10) Prove Theorem 4.6 "A language is context-free if and only if it is accepted by a PDA." Assume the grammar is in Greibach normal form.

(11) Construct a pushdown automaton that reads the same language as the grammar $\Gamma = (N, \Sigma, S, P)$ defined by $N = \{S, B\} \cup \Sigma$, $\Sigma = \{a, b, c\}$, and the set of productions P given by

$$S \to aA \qquad A \to aAB \qquad A \to a \qquad B \to b \qquad B \to \lambda.$$

(12) Construct a pushdown automaton that reads the same language as the grammar $\Gamma = (N, \Sigma, S, P)$ defined by $N = \{S, A, B\}$, $\Sigma = \{a, b, c\}$, and the set of productions P given by

$$S \to AB \qquad A \to abaA \qquad A \to \lambda \qquad B \to Bcacc \qquad B \to \lambda.$$

(13) Construct a pushdown automaton that reads the same language as the grammar $\Gamma = (N, \Upsilon, S, P)$ defined by $N = \{S, A, B\}$, $\Sigma = \{a, b, c\}$, and the set of productions P given by

$$S \to AdB \qquad A \to abaA \qquad A \to \lambda \qquad B \to Bcacb \qquad B \to \lambda.$$

(14) Construct a pushdown automaton that reads the same language as the grammar $\Gamma = (N, \Upsilon, S, P)$ defined by $N = \{S, A, B\}$, $\Upsilon = \{a, b, c\}$, and the set of productions P given by

$$S \to AB \qquad A \to acA \qquad B \to bcB \qquad B \to bB$$
$$A \to aAa \qquad B \to \lambda \qquad A \to \lambda.$$

(15) Construct a pushdown automaton that reads the same language as the grammar $\Gamma = (N, \Upsilon, S, P)$ defined by $N = \{S, A, B\}$, $\Upsilon = \{a, b, c, d\}$, and the set of productions P given by

$$S \to AB \qquad A \to aAc \qquad B \to bBc \qquad B \to bB$$
$$A \to AaA \qquad B \to \lambda \qquad A \to \lambda.$$

4.4   The Pumping Lemma and decidability

Just as we were able to show that there are languages that are not regular languages, we are also able to show that there are languages that are not context-free. We begin by returning to the concept of the parse tree or derivation tree. The **height** of the tree is the length of the longest path in the tree. The **level** of a vertex A in the tree is the length of the path from the vertex S to the vertex A. A tree is a **binary** tree if each vertex has at most two children. Note that if the grammar is in Chomsky normal form then every tree formed is a binary tree.

Lemma 4.13   *If $A \Rightarrow^* w$ where A is a nonterminal and the height of the corresponding derivation tree with root A is n, then the length of w is less than or equal to 2^{n-1}.*

Proof   We use induction on the height of the derivation tree. If $n = 1$, then the derivation has the form $A \to a$, and the length of $w = a$ is $1 = 2^0$. Assume the lemma is true when $n = k$, and let $A \Rightarrow^* w$ have a derivation tree of height $k + 1$. Then $A \Rightarrow BC \Rightarrow^* uv = w$, where $B \Rightarrow^* u$ and $C \Rightarrow^* v$ and the derivation tree for both of these derivations has height n. Therefore both u and v have length less than or equal to 2^{k-1} and w has length less than or equal to $2 \cdot 2^{k-1} = 2^k = 2^{(k+1)-1}$. □

Previously we had a pumping theorem for regular languages. We now have one for context-free languages.

Theorem 4.7 (Pumping Lemma) *Let L be a context-free language. There exists an integer M so that any word longer than M in L has the form $xuvwy$ where uw is not the empty word, the word uvw has length less than or equal to M, and $xu^n wv^n y \in L$ for all $n \geq 0$.*

Proof Let $L - \{\lambda\}$ be a nonempty language generated by the grammar $\Gamma = (N, \Sigma, S, P)$ in Chomsky normal form with p productions. Let $M = 2^p$. Assume there is a word w in L with length greater than or equal to M. Then by the previous theorem, the derivation tree has height greater than p. Therefore there is a path $S \to \cdots \to a$ where a is a letter in the derivation tree with length greater than p and a is a letter of w. Since there are only p productions, some nonterminal occurs more than once on the left-hand side of a production. Let C be the first nonterminal to occur the second time. Therefore we have a derivation

$$S \Rightarrow^* \alpha C \beta \Rightarrow^* xuCvy \Rightarrow^* xuwvy$$

where $\alpha \Rightarrow^* x$, $\beta \Rightarrow^* y$, $C \Rightarrow^* uCv$ and $C \Rightarrow w$. But using these derivations, we can form the derivation

$$S \Rightarrow xuCvy \Rightarrow^* xuuCvvy \Rightarrow^* xu^n wv^n y \text{ for any positive integer } n.$$

Since the first production in the derivation has the form $C \Rightarrow AB \Rightarrow^* uCv$, and there are no empty words, either u or v is not the empty word. Pick a letter a in uwv; we can work our way back to S using one occurrence of each of the productions $C \Rightarrow^* uCv$ and $C \Rightarrow w$. Hence the length of the path is at most p, and the length of uwv is less than or equal to M. □

We are now able to find a language which is not a context-free language.

Corollary 4.1 *The language $L = \{a^m b^m a^m : m \geq 1\}$ is not a context-free language.*

Proof Assume m is large enough so that the length of $a^m b^m a^m$ is larger than M. Therefore $a^m b^m a^m = puqvr$ where $pu^n qv^n r \in L$ for all $n \geq 1$. If either u or v contains both a and b, for example assume $u = a^i b^j$, then $(a^i b^j)^n$ must be a substring of $pu^n qv^n r$ which is clearly impossible. Thus u and v each consist entirely of strings of as or entirely of strings of bs. They cannot both be strings of as or strings of bs since the number of occurrences of the common letter in these strings could continue as n increases but the number of occurrences of the other letter would not increase. Therefore u must be a string of as to begin each word $a^m b^m a^m$ and v must be a string of as to end each word $a^m b^m a^m$ which is a contradiction. □

Example 4.19 The language $L = \{a^i b^j c^i d^j\} : i, j \geq 1\}$ is not a context-free language.

Proof Let M and $xu^n wv^n y$ be the same as in the previous theorem. Therefore $|uwv| \leq m$. Consider $a^m b^m c^m d^m \in L$. Since $|uwv| \leq m$ it must be contained in a power of one of the letters a, b, c, or d or in the powers of two adjacent letters. If it is contained in the power of one letter, say a, then there are fewer as than cs in each word in M, which is a contradiction. If it is contained in a power of two adjacent letters, say a and b, then there are fewer as than cs in each word in M, which is a contradiction. □

Example 4.20 The language $L = \{ww : w \in \{a, b\}^*\}$ is not context-free. We shall see in Theorem 4.10 that the intersection of a regular language and a context-free language is context-free. But $L \cap a^* b^* a^* b^* = \{a^i b^j a^i b^j\}$ is not context-free. The argument is the same as in the previous example.

Example 4.21 The language $L = \{x : \text{the length of } x \text{ is a prime}\}$ is not context-free. Since primes are arbitrarily large some element w with length m of L must have the form $xu^n wv^n y$ for $n \geq 2$. Let $m = |xwy|$. Then $|uv| = n - w$ and $|xu^m wv^m y| = m + m(n - w)$ which is not a prime.

We have previously seen that the set of regular languages is closed under the operations concatenation, union, Kleene star, intersection and complement. We now explore these same operations for the set of context-free languages.

Theorem 4.8 *The set of context-free languages is closed under the operations of concatenation, union, and Kleene star.*

Proof Let L_1 and L_2 be generated by grammars $\Gamma_1 = (N_1, \Sigma_1, S_1, P_1)$ and $\Gamma_2 = (N_2, \Sigma_2, S_2, P_2)$ respectively. Assume N_1 and N_2 are disjoint. This can always be accomplished by relabeling the elements in either N_1 or N_2.

The language $L_1 L_2$ can be generated by the grammar $\Gamma = (N, \Sigma, S, P)$ where $N = N_1 \cup N_2 \cup \{S\}$, $\Sigma = \Sigma_1 \cup \Sigma_2$, and $P = P_1 \cup P_2 \cup \{S \rightarrow S_1 S_2\}$. If $u \in L_1$ and $v \in L_2$, then $S_1 \Rightarrow^* u$, in Γ_1, $S_2 \Rightarrow^* v$ in Γ_2, and using leftmost derivation, we have $S \Rightarrow S_1 S_2 \Rightarrow^* u S_2 \Rightarrow^* uv$ in Γ.

The language $L_1 \cup L_2$ can be generated by the grammar $\Gamma = (N, \Sigma, S, P)$ where $N = N_1 \cup N_2 \cup \{S\}$, $\Sigma = \Sigma_1 \cup \Sigma_2$, and $P = P_1 \cup P_2 \cup \{S \rightarrow S_1, S \rightarrow$

S_2}. Let $w \in L_1 \cup L_2$. Therefore $w \in L_1$ or $w \in L_2$. If $w \in L_1$, then $S_1 \Rightarrow^* w$ in Γ_1, and $S \Rightarrow S_1 \Rightarrow^* w$ in Γ. If $w \in L_2$, then $S_2 \Rightarrow^* w$ in Γ_2, and $S \Rightarrow S_2 \Rightarrow^* w$ in Γ.

The language L_1^* can be generated by the grammar $\Gamma = (N, \Sigma, S, P)$ where $N = N_1 \cup \{S\}$, $\Sigma = \Sigma_1$ and $P = P_1 \cup P_2 \cup \{S \to S_1 S, S \to \lambda\}$. Let $w_1, w_2, w_3, \ldots, w_n \in L_1$. Using productions $S \to S_1 S$ and $S \to \lambda$, we can form the derivation $S \Rightarrow^* S_1^n = S_1 S_1 S_1 \cdots S_1$. Using leftmost derivations we can derive $S \Rightarrow^* S_1 S_1 S_1 \cdots S_1 \Rightarrow^* w_1 S_1 S_1 \cdots S_1 \Rightarrow^* w_1 w_2 S_1 \cdots S_1 \Rightarrow^* w_1 w_2 \cdots w_n$ in Γ. Hence $L_1^* = L$, the language generated by Γ. $\square$

Theorem 4.9 *The set of context-free languages is not closed under the operations of intersection and complement.*

Proof The sets $\{a^n b^n c^m : m, n \geq 0\}$ and $\{a^n b^m c^m : m, n \geq 0\}$ are context-free. The first is generated by the grammar with productions

$$P = \{S \to BC, B \to aBb, B \to \lambda, C \to cC, C \to \lambda\}.$$

The second is generated by the grammar with productions

$$P = \{S \to AB, A \to aA, A \to \lambda, B \to bBc, B \to \lambda\}.$$

However, the intersection is the language $L = \{a^m b^m a^m : m \geq 0\}$, which we have shown is not context-free.

If the set of context-free languages is closed under complement then since $L_1 \cap L_2 = (L_1' \cup L_2')'$, the set of context-free languages is closed under intersection which we have already shown is not true. $\square$

Although the intersection of context-free languages is not necessarily a context-free language, the intersection of a context-free language and a regular language is a context-free language. The proof is somewhat similar to the one showing that the union of languages accepted by an automaton is accepted by an automaton.

Theorem 4.10 *The intersection of a regular language and a context-free language is context-free.*

Proof Let the pushdown automaton $M = (\Sigma, Q, s, I, \Upsilon, F)$ where Σ is the alphabet, Q is the set of states, s is the initial or starting state, I is the set of stack symbols, F is the set of acceptance states, and Υ is the transition relation where the relation Υ is a finite subset of $((Q \times \Sigma^* \times I^*) \times (Q \times I^*))$. Let the deterministic finite automaton $M_1 = (\Sigma_1, Q_1, q_0, \Upsilon_1, F_1)$ where Σ_1 is the alphabet, Q_1 is the set of states, q_0 is the initial or starting state, F_1 is the set of acceptance states, and Υ_1 is the transition function. We now define the pushdown automaton

$M_2 = (\Sigma_2, Q_2, s_2, I_2, \Upsilon, F_2)$ where $s_2 = (s, q_0)$, $\Sigma_2 = \Sigma_1 \cup \Sigma$, $I_2 = I$, $Q_2 = Q \times Q_1$, and $F_2 = F \times F_1$. Define Υ_2 by $(((s_i, q_j), a, X), ((s_m, q_n), b)) \in \Upsilon_2$ if and only if $((s_i, a, X), \vdash (s_m, b))$ in M and $\Upsilon_1(q_j, u) = q_n$ in M_1. A word is w accepted in M_2 if and only if $((s, q_0), w, \lambda) \vdash_2^* ((s_a, q_b), \lambda)$ in M_2, where s_a and q_b are acceptance states in M and M_1 respectively. Thus w is accepted by the pushdown automaton M and also accepted by M_1.

To show $M_2(L) = M(L) \cap M_1(L)$, the reader is asked to first show that $((s, q_0), w, \lambda) \vdash^* ((s_m, q_n), \lambda, \alpha)$ if and only if $(s, w, \lambda) \vdash^* (s_m, \lambda, \alpha)$ and $(q_0, w) \vdash_1^* (q_n, \lambda)$, using induction on the number of operations in $\vdash_2^*$. The theorem immediately follows. □

Definition 4.10 *A nonterminal in a context-free grammar is **useless** if it does not occur in any derivation $S \Rightarrow^* w$, for $w \in \Sigma^*$. If a nonterminal is not useless, then it is **useful**.*

Theorem 4.11 *Given a context-free grammar, it is possible to find and remove all productions with useless nonterminals.*

Proof We first remove any nonterminal U so that there is no derivation $U \Rightarrow^* w$, for $w \in \Sigma^*$. To find such nonterminals, let X be defined as follows: (1) For each nonterminal V such that $V \to w$ is a production for $w \in \Sigma^*$, let $V \in X$. (2) If $V \to V_1 V_2 \cdots V_n$ where $V_i \in X$ or Σ^* for $1 \le i \le n$, let $V \in X$. Continue step (2) until no new nonterminals are added to X. Any nonterminal U not in X has no derivation $U \Rightarrow^* w$. If S is not in X, then the language generated by the context-free grammar is empty and we are done. There are no useful nonterminals. Assume S is in X. All productions containing nonterminals not in X are removed from the set of productions P.

Assume such productions have been removed. We now have to remove any nonterminal U which is not reachable by S, i.e. there is no production $S \Rightarrow^* W$ where U is in the string W. To test each nonterminal U we form a set Y_U as follows: (1) If $V \to W$ and U is in the string W, then $V \in Y_U$. (2) if $R \Rightarrow T$, where an element of Y_U is in the string T, then $R \in Y_U$. Continue step (2) until no new nonterminals are added to Y_U. If $S \in Y_U$, then U is reachable by S. If not, U is not reachable by S. Remove all productions which contain nonterminals not reachable by S. The context-free grammar created contains no useless nonterminals. □

Theorem 4.12 *It is possible to determine whether a context-free language L is empty.*

Proof A context-free language L is empty if and only if it contains no useful nonterminals. □

Theorem 4.13 *Given $w \in \Sigma^*$, and a context-free grammar G, it is possible to determine whether w is in the language generated by G.*

Proof Assume that G is in Chomsky normal form. Let w be a word of length $n \geq 1$. Each step of the derived string becomes longer except when the nonterminal is replaced by a terminal. Therefore the derivation $S \Rightarrow^* w$ has length $k \leq 2^{n-1}$ since G is in Chomsky normal form. Therefore check all derivations of length $k \leq 2^{n-1}$. $\square$

The above proof shows that it is possible to determine whether a word is in the language generated by a grammar; it really is not a practical way.

Theorem 4.14 *Let G be a context-free grammar in Chomsky normal form with exactly p productions. The language $L(G)$ is infinite if and only if there exists a word ω in $L(G)$ such that $2^p < |\omega| < 2^{p+1}$.*

Proof If there is a word with length greater than 2^p then by the proof of the Pumping Lemma, $L(G)$ is infinite. Conversely, let ω be the shortest word with length greater than 2^{p+1}. By the Pumping Lemma, $\omega = xu^i wv^i y$, where the length of $uvw \leq 2^p$ and $\mu = xu^{i-1}wv^{i-1}y$ is in $L(G)$. But $|\mu| > |\omega| - |uv| \geq 2^p$. Also $|\mu| < |\omega|$ and w is the shortest word with length greater than or equal to 2^{p+1}. Therefore $|\mu| < 2^{p+1}$. $\square$

Theorem 4.15 *It is possible to determine whether a language generated by a context-free grammar is finite or infinite.*

Proof Since it is possible to determine whether a word is in the language of a context-free grammar, simply try all words with length between 2^p and 2^{p+1} to see if one of them is in the context-free grammar. If one is, the grammar is infinite. If not the grammar is finite. $\square$

Exercises

(1) Let grammar $\Gamma = (N, \Upsilon, S, P)$ be defined by $N = \{S, A, B\}$, $\Upsilon = \{a, b, c\}$, and the set of productions P given by

$$S \to AB \qquad A \to acA \qquad B \to bcB \qquad B \to bB$$
$$A \to aBa \qquad B \to \lambda \qquad A \to a.$$

Let L be the language generated by Γ. Find the grammar that generates L^*.

(2) Let L_1 be the language generated by the grammar $\Gamma_1 = (N, \Sigma, S, P)$ defined by $N = \{S, A, B\}$, $\Sigma = \{a, b, c\}$, and the set of productions P

given by

$$S \to aA \qquad A \to aAB \qquad A \to a \qquad B \to b \qquad B \to \lambda,$$

and L_2 be the language generated by the grammar $\Gamma_2 = (N, \Sigma, S, P)$ defined by $N = \{S, A, B\}$, $\Sigma = \{a, b, c\}$, and the set of productions P given by

$$S \to AB \qquad A \to abaA \qquad A \to \lambda \qquad B \to Bcacc \qquad B \to \lambda.$$

Find the grammar that generates $L_1 L_2$.

(3) Let L_1 be the language generated by the grammar $\Gamma_1 = (N, \Upsilon, S, P)$ defined by $N = \{S, A, B\}$, $\Sigma = \{a, b, c\}$, and the set of productions P given by

$$S \to AdB \qquad A \to abaA \qquad A \to \lambda \qquad B \to Bcacb \qquad B \to \lambda,$$

and L_2 be the language generated by the grammar $\Gamma_2 = (N, \Upsilon, S, P)$ defined by $N = \{S, A, B\}$, $\Upsilon = \{a, b, c\}$, and the set of productions P given by

$$S \to AB \qquad A \to acA \qquad B \to bcB \qquad B \to bB$$
$$A \to aAa \qquad B \to \lambda \qquad A \to \lambda.$$

Find the grammar that generates $L_1 \cup L_2$.

Determine whether the following languages are context-free. If the language is context-free, construct a grammar that generates it. If it is not context-free, prove that it is not.

(4) $L = \{a^m b^n c^{2n} : m, n = 1, 2, \ldots\}$.

(5) $L = \{ww^R w : w \in \{a, b\}^*\}$.

(6) $L = \{w \in \{a, b, c\}^*\} : w$ has an equal number of as and bs $\}$.

(7) $L = \{a^n b^{2n} c^n : n = 1, 2, \ldots\}$.

(8) Prove the induction step in Theorem 4.10.

5

Turing machines

5.1 Deterministic Turing machines

The Turing machine is certainly the most powerful of the machines that we have considered and, in a sense, is the most powerful machine that we can consider. It is believed that every well-defined algorithm that people can be taught to perform or that can be performed by any computer can be performed on a Turing machine. This is essentially the statement made by Alonzo Church in 1936 and is known as Church's Thesis. This is not a theorem. It has not been mathematically proven. However, no one has found any reason for doubting it.

It is interesting that although the computer, as we know it, had not yet been invented when the Turing machine was created, the Turing machine contains the theory on which computers are based. Many students have been amazed to find that, using a Turing machine, they are actually writing computer programs. Thus computer programs preceded the computer.

We warn the reader in advance that if they look at different books on Turing machines, they will find the descriptions to be quite different. One author will state a certain property to be required of their machine. Another author will strictly prohibit the same property on their machine. Nevertheless, the machines, although different, have the same capabilities.

The Turing machine has an input alphabet Σ, a set of tape symbols, Γ containing Σ, and a set of states Q, similar to the automaton. The Turing machine has two special states, the start state s_0 and the halt state h. When the machine reaches the halt state it shuts down. It also has a tape which is infinitely long on the right. If made of paper it can wipe out a forest.

The tape contains squares on which letters of the alphabet and other symbols can be written or erased. Only a finite number of the squares may contain tape symbols. All of the squares to the right of the last square containing a tape symbol are considered to be blank. Some of the squares between or in front of letters may also be blank. In addition to the tape there is a head which can read any tape symbol which is on the square of the tape at which the head is pointing. It also can be in different states just like an automaton and a pushdown automaton. Also like the automaton, the input can be a letter of the alphabet which can be read from the tape together with the current state of the machine. Depending on the input and its current state, the machine, in addition to changing states, can print a different symbol on the square of the tape in front of it or erase the letter in the square. In addition or instead, the head can move left or right from the square it has just read to the next square. The blank can be both read and printed by the Turing machine, but is not considered an element of the tape symbols. As input, reading a blank is simply reading the absence of any of the tape symbols. Printing a blank is considered to be erasing the symbol currently in that square. We use # for blank. The Turing machine shown below is in state s_1 and is reading letter a.

More formally we have the following definition.

Definition 5.1   *A **deterministic** Turing machine is a quintuple*

$$(Q, \Sigma, \Gamma, \delta, s_0, h)$$

where Q is the set of states, Γ is a finite set of tape symbols, which includes the alphabet and #, s_0 is the starting state, h is the halt state, and δ is a function from $Q \times \Gamma$ to $Q \times \Gamma \times N$ where N consists of L which indicates a movement on the tape one position to the left, R which indicates a movement on the tape one position to the right, and # which indicates that no movement takes place.

Just like any computer, a Turing machine has a program or set of rules which tell the machine what to do. An example of a rule is

$$\delta(s_1, a) = (s_2, b, L)$$

which we shall denote as

$$(s_1, a, s_2, b, L).$$

This rules says that if the machine is in state s_1 and reads the letter a, it is to change to state s_2, print the letter b in place of the letter a and move one square to the left. The rule

$$(s_1, a, s_2, \#, R)$$

says that if the machine is in state s_1 and reads the letter a, it changes to state s_2, erases the a and moves one square to the right. The rule

$$(s_1, \#, h, \#, \#)$$

says that if the machine is in state s_1 and reads a blank then it halts, and thus does not print anything or move the position on the tape. For consistency we shall always require that the machine begins in the leftmost square.

It may appear that since δ is a function, the deterministic Turing machine will either continue forever or reach the halt state. However, if the Turing machine is reading the leftmost square on the tape and gets the command to move left, it obviously cannot do so. In such a case we say that the system **crashes**. Often a special symbol is placed in the first box to warn the machine that it is reaching the end of the tape.

Obviously there is a difference between the machine stopping because it crashes and stopping when it reaches the halt state. In the second case the machine has completed its program.

It is obvious that our rules allow us both to print a letter and move the position on the tape to the left or to the right. Some definitions allow a machine either to print a letter or to move the head, but not both. Thus it requires two separate rules to print a letter and move the position on the tape.

We shall begin with a program that simply moves the position of the machine on the tape from the beginning to the end of a string. The alphabet is $\Sigma = \{a, b\}$ and symbols $\Gamma = \{a, b, \#\}$. We shall have the set of states $Q = \{s_0, s_1, h\}$ and the set of rules

$$(s_0, a, s_1, a, R) \qquad (s_0, b, s_1, b, R) \qquad (s_1, a, s_1, a, R)$$
$$(s_1, b, s_1, b, R) \qquad (s_1, \#, h, \#, \#,).$$

This program leaves everything alone. It simply reads each letter and then moves right to the next square. When it reaches a blank, it shuts down. However, this program does do something which we shall later need. It moves the position on the tape from the beginning of the word to the end of the word. Instead of having it reach a blank and shut down, we will put it at the beginning of another program where we want the position of the machine to be at the end of the word. Hence we shall call this program **go-end**. As we demonstrate this program, it would be rather tiresome to continually draw the Turing machine so rather than draw

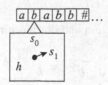

which shows the position of the machine at the second square of the tape and in state s_1, while the first and third squares of the tape contain an a, the second, fourth and fifth squares contain a b and the other squares are blank; we replace this with

$$1$$
$$a \quad \underline{b} \quad a \quad b \quad b$$

where the line below the b denotes the location of the head, and the 1 above the b denotes the current state of the machine. We shall call this the **configuration** of the Turing machine.

As we begin our program the machine has configuration

$$0$$
$$\underline{a} \quad b \quad a \quad b \quad b.$$

We then apply rule

$$(s_0, a, s_1, a, R)$$

moving the head to the right and changing from state s_0 to state s_1 and our machine then has configuration

$$1$$
$$a \quad \underline{b} \quad a \quad b \quad b.$$

We then apply rule

$$(s_1, b, s_1, b, R)$$

moving the head to the right again and our machine then has configuration

$$1$$
$$a \quad b \quad \underline{a} \quad b \quad b.$$

We then apply rule

$$(s_1, a, s_1, a, R)$$

moving the head to the right again and our machine then has configuration

$$1$$
$$a \quad b \quad a \quad \underline{b} \quad b.$$

We again apply rule

$$(s_1, a, s_1, a, R)$$

moving the head to the right again and our machine then has configuration

$$1$$
$$a \quad b \quad a \quad b \quad \underline{b}.$$

We apply the same rule again and have

$$
\begin{matrix}
& & & & 1 & \\
a & b & a & b & b & \underline{\#}.
\end{matrix}
$$

We then use rule

$$(s_1, \#, h, \#, \#,)$$

and the machine shuts down.

We mentioned previously that if the position on the tape is on the leftmost position on the tape and gets an instruction to move left, we say that the machine **crashes**, and the machine ceases functioning.

We shall now construct a rather unusual program. This program causes the machine to crash. We shall again let the input alphabet Γ be the set $\{a, b, \#\}$. We shall also assume we have states $Q = \{s_0, s_1, \ldots, s_j, \ldots\}$. It shall have the rules

$$(s_j, a, s_j, a, L) \qquad (s_j, b, s_j, b, L) \qquad (s_j, \#, s_j, \#, L).$$

If we have a larger alphabet, we simply add more rules, so that regardless of what the machine reads when it is in state s_j, it continues to go left until it crashes. This program does not begin at s_0 because we want to include it in other programs when we want to crash the system. We shall call this program **go-crash**. State s_j is the "suicide" state. When we want to crash the system we simply instruct it to go to state s_j.

It seems pretty silly to think of either *go-end* or *go-crash* as complete programs. We really want to use them inside other programs. We shall refer to these types of program as **subroutines**.

The reason for the go-crash program is really theoretical. If δ is a partial function instead of a function then the Turing machine is still deterministic in the sense that for every input for which there is a rule, there is a unique output. If for *every* input, there is a unique output, then the set of rules would define a function. If the rules do not define a function then there is a state s and an input letter a for which there is no rule. When this happens, we say that the system **hangs**, since it cannot go on. We shall again meet this problem with nondeterministic automata. Suppose we would like the set of rules to define a function, but we still want the program to stop when it is in state s and reads a. The system cannot hang since the function is defined for every input. We can however add a rule

$$(s, a, s_j, a, L)$$

which puts the system into the suicide state and causes it to crash using go-crash. Thus the system crashes instead of hanging and we have expanded our rules so that we have a function. In this discussion, we will state only relevant rules with the understanding that we could produce a function using go-crash if we really wanted to do so.

It is perhaps time we considered something a bit more practical for the Turing machine. We begin by showing some of its properties as a text editor. Our first step is not exactly a giant one. We show how to move the position on the tape to the right n steps. Again we assume both the input and output alphabet are the set $\{a, b\}$. If we have a larger alphabet, we simply add appropriate rules for each new letter. The set of states $Q = \{s_1, \ldots, s_j, \ldots, s_n, s_{n+1}\}$. We shall call this new subroutine **go-right**(n). It has the following rules:

$$(s_1, a, s_2, a, R) \quad (s_2, a, s_3, a, R) \quad (s_3, a, s_4, a, R) \cdots (s_n, a, s_{n+1}, a, R)$$
$$(s_1, b, s_2, b, R) \quad (s_2, b, s_3, b, R) \quad (s_3, b, s_4, b, R) \cdots (s_n, b, s_{n+1}, b, R).$$

It is easily seen that if we begin in state s_1, each application of a rule, regardless of the letter read, moves the the position on the tape one step to the right and increases the state. After n steps the head has been moved to the right by n squares and we are in state s_{n+1}. It is hoped that, with little effort, the reader can create a subroutine for moving to the left by n squares.

Suppose that after moving left or right by n squares, or without moving at all we want to change the letter in the current square occupied from a to b. Assuming that we are in state s_i at the time then we simply use the rule

$$(s_i, a, s_i, b, \#).$$

Moving along, suppose that

$$\Sigma = \{a_1, a_2, a_3, \ldots, a_n, b_1, b_2, b_3, \ldots, b_n\},$$

$\Gamma = \Sigma \cup \{\#\}$ and we want to replace

$$a_1 a_2 \ldots a_i, a_{i+1} \ldots a_j, \ldots a_n$$

with

$$a_1 a_2 \ldots a_i, b_{i+1} \ldots b_j, a_{j+1} \ldots a_n.$$

We first use *go-right*(i) to move to the proper position so the head is on a_{i+1}. Assume we are in state s', We then use the rules

$$(s', a_{i+1}, s_1', b_{i+1}, R)$$
$$(s_1', a_{i+2}, s_2', b_{i+2}, R)$$
$$(s_2', a_{i+3}, s_3', b_{i+3}, R)$$
$$(s_3', a_{i+4}, s_4', b_{i+4}, R)$$
$$\vdots$$
$$(s_{j-i-1}', a_j, s_{j-i}', b_j, R)$$

to replace the letters and use *go-left*(j) to return to the original spot.

The next text edit feature which we shall illustrate is to insert a letter in a string. We shall find this feature very handy in the near future. We shall call this subroutine **insert(c)**. Say that we have a string

$$a_1 a_2 \cdots a_i a_{i+1} \cdots a_{n-1} a_n$$

and we want to replace it with

$$a_1 a_2 \cdots a_i c a_{i+1} \cdots a_{n-1} a_n$$

so that the string $a_{i+1} \cdots a_{n-1} a_n$ must be moved one square to the right and c placed in the square formerly occupied by a_{i+1}. We shall assume that the string contains no blanks. If it does, a special symbol will have to be used to denote the end of the string. Actually this is not quite the order in which we shall proceed. First, for simplicity, assume that the input and output alphabets are the same and that $\Sigma = \{a, b, c\}$ and $\Gamma\{a, b, c, \#\}$. We shall assume that we know the position on the tape in which the c is to be placed (i.e. we know i) and that we know the length of the string (i.e. we know n). First we use *go-right(i)* to place the head where the letter c is to be placed. Assume that we are in state s_x when we reach this square. We are going to need a state for each letter in the alphabet. Thus we shall need s_a, s_b, and s_c. The process is really rather simple. When we print c, we need to remember a_{i+1} so that we can print it in the next square. We do this by entering $s_{a_{i+1}}$ after we have printed c and then moving right. In state $s_{a_{i+1}}$, we print a_{i+1} in the square occupied by a_{i+2} and then enter state $s_{a_{i+2}}$ and again move right. Each time we print a letter, we enter the state corresponding to the letter destroyed and in this way "remember" this letter so it can be printed in the next square. Remember in state $s_{a_{i+j}}$, we print a_{i+j} regardless of the letter read. Finally, when we reach a blank square, we print a_n and then use *go-left(n)* to return to the beginning of the string. Also it is possible that c occurs elsewhere in the string; however, we shall assume that a_{i+1} is not already c. Thus our rules for actually printing c and moving over the other letters are

(s_x, a, s_a, c, R)	(s_b, c, s_c, b, R)	(s_x, b, s_b, c, R)	(s_c, a, s_a, c, R)
(s_c, b, s_b, c, R)	(s_a, b, s_b, a, R)	(s_c, c, s_c, c, R)	(s_a, c, s_c, a, R)
(s_b, a, s_a, b, R)	$(s_b, \#, s_y, b, \#)$	(s_b, b, s_b, b, R)	$(s_c, \#, s_y, c, \#)$
(s_a, a, s_a, a, R)	$(s_a, \#, s_y, a, \#)$		

and we end up in state s_y.

For example assume we have the word *abbac* and want to insert c so that we have *abcbbc*. Using *go-right(2)*, we have configuration

$$x$$
$$a \quad b \quad \underline{b} \quad a \quad c.$$

Applying rule

$$(s_x, b, s_b, c, R)$$

we have configuration

$$\begin{array}{ccccc} & & & b & \\ a & b & c & \underline{a} & c. \end{array}$$

In the future we will condense this statement to

$$(s_x, b, s_b, c, R) \vdash \begin{array}{ccccc} & & & b & \\ a & b & c & \underline{a} & c. \end{array}$$

We then have the following rules and configurations

$$(s_b, a, s_a, b, R) \vdash \begin{array}{ccccc} & & & & a \\ a & b & c & b & \underline{c} \end{array}$$

$$(s_a, c, s_c, a, R) \vdash \begin{array}{cccccc} & & & & & c \\ a & b & c & b & a & \underline{\#} \end{array}$$

$$(s_a, \#, s_y, a, \#) \vdash \begin{array}{cccccc} & & & & & y \\ a & b & c & b & a & \underline{c} \end{array}$$

and we now use *go-left*(5) to return to our original position.

Suppose we began at the square of the letter we were replacing and wanted to return to that square. Instead of placing the c in the square, we would place a marker, and then when we had finished moving the letters we would return to the marker and replace it with a c. Details are left to the reader.

The next text edit feature which we shall illustrate is to delete a letter in a string and close up the empty square. We shall call this subroutine **delete**(c). Say that we have a string $a_1 a_2 \cdots a_i c a_{i+1} \cdots a_{n-1} a_n$ which contains no blanks and we want to replace it with $a_1 a_2 \cdots a_i a_{i+1} \cdots a_{n-1} a_n$. If there is a blank, we would have to have a special marker to denote the end of the string. There are at least two ways of doing this. One way is to move over to the square containing c and replace it with a marker which is not part of the regular alphabet. Then go to the end of the string and move each letter to the left in a similar manner to the one we used to move letters to the right in *insert*(c), replacing the marker with the letter to its right and then changing states to return to the front of the word or wherever desired. If the string contains no blanks then a blank can be used to denote the end of the string. Otherwise a special symbol will need to be used. The details of this subroutine are left to the reader.

An alternative form is to move to the letter to be deleted, replace it with a marker, move to the right to find the next letter and replace the marker with that letter. Then go right again to the letter which has been duplicated and replace it with a marker. Continue this process until reaching the end of the string. Let Δ be the special marker. Assume that we have used *go-right(i)* to reach the letter c to be deleted. Again assume that we begin in state s_x and want to end in state s_y. We shall also let the $\Sigma = \{a, b, c\}$ and $\Gamma = \{a, b, c\} \cup \{\#, \}$. We then have the following set of rules:

$(s_x, c, s_{x'}, \Delta, R)$ $(s_x, b, s_{x'}, \Delta, R)$ $(s_x, a, s_{x'}, \Delta, R)$ $(s_{x'}, a, s_a, a, L)$

$(s_{x'}, b, s_b, b, L)$ $(s_{x'}, c, s_c, c, L)$ (s_a, Δ, s_x, a, R) (s_b, Δ, s_x, b, R)

(s_c, Δ, s_x, c, R) $(s_{x'}, \#, s_{x''}, \#, L)$ $(s_{x''}, \Delta, s_y, \#, \#)$.

Note that the marker is not actually needed. It is used to make the rules easier to read.

For example, suppose we have the string *abcbac* and wish to remove the c in the third space. We use *go-right*(2) to get to the desired space and have the configuration

$$
\begin{array}{cccccc}
 & & x & & & \\
a & b & \underline{c} & b & a & c.
\end{array}
$$

We then have the following rules and configurations:

$(s_x, c, s_{x'}, \Delta, R) \vdash$
$$
\begin{array}{cccccc}
 & & & x' & & \\
a & b & \Delta & \underline{b} & a & c
\end{array}
$$

$\Rightarrow (s_{x'}, b, s_b, b, L) \vdash$
$$
\begin{array}{cccccc}
 & & b & & & \\
a & b & \underline{\Delta} & b & a & c
\end{array}
$$

$\Rightarrow (s_b, \Delta, s_x, b, R) \vdash$
$$
\begin{array}{cccccc}
 & & & x & & \\
a & b & b & \underline{b} & a & c
\end{array}
$$

$\Rightarrow (s_x, b, s_{x'}, \Delta, R) \vdash$
$$
\begin{array}{cccccc}
 & & & & x' & \\
a & b & b & \Delta & \underline{a} & c
\end{array}
$$

$\Rightarrow (s_{x'}, a, s_a, a, L) \vdash$
$$
\begin{array}{cccccc}
 & & & a & & \\
a & b & b & \underline{\Delta} & a & c
\end{array}
$$

$\Rightarrow (s_a, \Delta, s_x, a, R) \vdash$
$$
\begin{array}{cccccc}
 & & & & x & \\
a & b & b & a & \underline{a} & .c
\end{array}
$$

$\Rightarrow (s_x, a, s_{x'}, \Delta, R) \vdash$
$$
\begin{array}{cccccc}
 & & & & & x' \\
a & b & b & a & \Delta & \underline{c}
\end{array}
$$

$\Rightarrow (s_{x'}, c, s_c, c, L) \vdash$
$$
\begin{array}{cccccc}
 & & & & c & \\
a & b & b & a & \underline{\Delta} & c
\end{array}
$$

$\Rightarrow (s_c, \Delta, s_x, c, R) \vdash$
$$
\begin{array}{cccccc}
 & & & & & x \\
a & b & b & a & c & \underline{c}
\end{array}
$$

$\Rightarrow (s_x, c, s_{x'}, \Delta, R) \vdash$
$$
\begin{array}{ccccccc}
 & & & & & & x' \\
a & b & b & a & c & \Delta & \underline{\#}
\end{array}
$$

$\Rightarrow (s_{x'}, \#, s_{x''}, \#, L) \vdash$
$$
\begin{array}{ccccccc}
 & & & & & x'' & \\
a & b & b & a & c & \underline{\Delta} & \#.
\end{array}
$$

Finally applying rule

$$(s_{x''}, \Delta, s_y, \#, \#)$$

we have configuration

$$
\begin{array}{ccccccc}
 & & & & & y & \\
a & b & b & a & c & \underline{\#} & \#
\end{array}
$$

and using *go-left*(5), we return to the beginning of the string.

Finally we show how to use the Turing machine to duplicate a string. For simplicity we shall limit the letters in the string to the set $\{a, b\}$. If the alphabet is increased, similar rules to those given will be added for each letter included. We shall need additional symbols λ_a, λ_b, σ_a, and σ_b. Briefly the first letter of the string is replaced by λ_a if the letter is a and by λ_b if the letter is b. We then go to the end of the string and place a corresponding σ_a, or σ_b. We then return to the first symbol and replace it with the original letter, go to the second letter and repeat the process. We continue until we have a string followed by corresponding σ_as, and σ_bs. We then replace each σ_a with an a and σ_b with a b.

Assume that we start in state s_x and end in state s_y. We then have the following set of rules:

$$(s_x, a, s_a, \lambda_a, R) \quad (s_b, \#, s_{x'}, \sigma_b, L) \quad (s_a, a, s_a, a, R) \quad (s_{x'}, a, s_{x'}, a, L)$$

$$(s_{x'}, b, s_{x'}, b, L) \quad (s_a, \sigma_a, s_a, \sigma_a, R) \quad (s_{x'}, \sigma_a, s_{x'}, \sigma_a, L) \quad (s_a, \sigma_b, s_a, \sigma_b, R)$$

$$(s_x, b, s_b, \lambda_b, R) \quad (s_{x'}, \lambda_a, s_x, a, R) \quad (s_b, a, s_b, a, R) \quad (s_{x'}, \lambda_b, s_x, b, R)$$

$$(s_x, \sigma_a, s_x, a, R) \quad (s_b, \sigma_a, s_b, \sigma_a, R) \quad (s_x, \sigma_b, s_x, b, R) \quad (s_b, \sigma_b, s_b, \sigma_b, R)$$

$$(s_a, \#, s_{x'}, \sigma_a, L) \quad (s_a, b, s_a, b, R) \quad (s_{x'}, \sigma_b, s_{x'}, \sigma_b, L) \quad (s_b, b, s_b, b, R)$$

$$(s_x, \#, s_y, \#, \#).$$

For example, we shall duplicate the word *bab*. The initial configuration is

$$
\begin{array}{ccc}
x & & \\
\underline{b} & a & b.
\end{array}
$$

We then have the following rules and configurations:

$$
(s_x, b, s_b, \lambda_b, R) \vdash \quad
\begin{array}{ccc}
 & b & \\
\lambda_b & \underline{a} & b
\end{array}
$$

$$
\Rightarrow (s_b, a, s_b, a, R) \vdash \quad
\begin{array}{ccc}
 & & b \\
\lambda_b & a & \underline{b}
\end{array}
$$

$$
\Rightarrow (s_b, b, s_b, b, R) \vdash \quad
\begin{array}{cccc}
 & & & b \\
\lambda_b & a & b & \underline{\#}
\end{array}
$$

$$
\Rightarrow (s_b, \#, s_{x'}, \sigma_b, L) \vdash \quad
\begin{array}{cccc}
 & & x' & \\
\lambda_b & a & \underline{b} & \sigma_b
\end{array}
$$

$$
\Rightarrow (s_{x'}, b, s_{x'}, b, L) \vdash \quad
\begin{array}{cccc}
 & x' & & \\
\lambda_b & \underline{a} & b & \sigma_b
\end{array}
$$

$$\Rightarrow (s_{x'}, a, s_{x'}, a, L) \vdash \quad \underset{\underline{\lambda_b} \quad a \quad b \quad \sigma_b}{\overset{x'}{}}$$

$$\Rightarrow (s_{x'}, \lambda_b, s_x, b, R) \vdash \quad \underset{b \quad \underline{a} \quad b \quad \sigma_b}{\overset{x}{}}$$

$$\Rightarrow (s_x, a, s_a, \lambda_a, R) \vdash \quad \underset{b \quad \lambda_a \quad \underline{b} \quad \sigma_b}{\overset{a}{}}$$

$$\Rightarrow (s_a, b, s_a, b, R) \vdash \quad \underset{b \quad \lambda_a \quad b \quad \underline{\sigma_b}}{\overset{a}{}}$$

$$\Rightarrow (s_a, \sigma_b, s_a, \sigma_b, R) \vdash \quad \underset{b \quad \lambda_a \quad b \quad \sigma_b \quad \underline{\#}}{\overset{a}{}}$$

$$\Rightarrow (s_a, \#, s_{x'}, \sigma_a, L) \vdash \quad \underset{b \quad \lambda_a \quad b \quad \underline{\sigma_b} \quad \sigma_a}{\overset{x'}{}}$$

$$\Rightarrow (s_{x'}, \sigma_b, s_{x'}, \sigma_b, L) \vdash \quad \underset{b \quad \lambda_a \quad \underline{b} \quad \sigma_b \quad \sigma_a}{\overset{x'}{}}$$

$$\Rightarrow (s_{x'}, b, s_{x'}, b, L) \vdash \quad \underset{b \quad \underline{\lambda_a} \quad b \quad \sigma_b \quad \sigma_a}{\overset{x'}{}}$$

$$\Rightarrow (s_{x'}, \lambda_a, s_x, a, R) \vdash \quad \underset{b \quad a \quad \underline{b} \quad \sigma_b \quad \sigma_a}{\overset{x}{}}$$

$$\Rightarrow (s_x, b, s_b, \lambda_b, R) \vdash \quad \underset{b \quad a \quad \lambda_b \quad \underline{\sigma_b} \quad \sigma_a}{\overset{b}{}}$$

$$\Rightarrow (s_b, \sigma_b, s_b, \sigma_b, R) \vdash \quad \underset{b \quad a \quad \lambda_b \quad \sigma_b \quad \underline{\sigma_a}}{\overset{b}{}}$$

$$\Rightarrow (s_b, \sigma_b, s_b, \sigma_b, R) \vdash \quad \underset{b \quad a \quad \lambda_b \quad \sigma_b \quad \underline{\sigma_a}}{\overset{b}{}}$$

$$\Rightarrow (s_b, \sigma_a, s_b, \sigma_a, R) \vdash \quad \underset{b \quad a \quad \lambda_b \quad \sigma_b \quad \sigma_a \quad \underline{\#}}{\overset{b}{}}$$

$$\Rightarrow (s_b, \#, s_{x'}, \sigma_b, L) \vdash \quad \underset{b \quad a \quad \lambda_b \quad \sigma_b \quad \underline{\sigma_a} \quad \sigma_b}{\overset{x'}{}}$$

$$\Rightarrow (s_{x'}, \sigma_a, s_{x'}, \sigma_a, L) \vdash \quad \underset{b \quad a \quad \lambda_b \quad \underline{\sigma_b} \quad \sigma_a \quad \sigma_b}{\overset{x'}{}}$$

$$\Rightarrow (s_{x'}, \sigma_b, s_{x'}, \sigma_b, L) \vdash \quad \underset{b \quad a \quad \underline{\lambda_b} \quad \sigma_b \quad \sigma_a \quad \sigma_b}{\overset{x'}{}}$$

$$\Rightarrow (s_{x'}, \lambda_b, s_x, b, R) \vdash \quad \underset{b \quad a \quad b \quad \underline{\sigma_b} \quad \sigma_a \quad \sigma_b}{\overset{x}{}}$$

$$\Rightarrow (s_x, \sigma_b, s_x, b, R) \vdash \quad \underset{b \quad a \quad b \quad b \quad \underline{\sigma_a} \quad \sigma_b}{\overset{x}{}}$$

$$\Rightarrow (s_x, \sigma_a, s_x, a, R) \vdash \quad \underset{b \quad a \quad b \quad b \quad a \quad \underline{\sigma_b}}{\overset{x}{}}$$

$$\Rightarrow (s_x, \sigma_b, s_x, b, R) \vdash \quad \underset{b \quad a \quad b \quad b \quad a \quad b \quad \underline{\#}}{\overset{x}{}}$$

and applying rule

$$(s_x, \#, s_y, \#, \#)$$

we have

$$
\overset{y}{\underset{}{\underline{\#}}}
$$

$$b \quad a \quad b \quad b \quad a \quad b \quad \overset{y}{\underline{\#}}$$

and we are done.

Definition 5.2 *A word $w \in \Sigma^*$ is **accepted** by a Turing machine T if, beginning in the start state, there is a way to read w and be in the halt state. The language accepted by a Turing machine T is the set of all words accepted by T.*

We next show how to use the machine as an acceptor. We begin by showing that a Turing machine can recognize a regular language. We already know that an automaton recognizes a regular language, so what we shall basically do is program it to imitate an automaton. Assume that we have a word in the Turing machine which we want the machine to read so that it can determine whether it wants to accept it. An automaton reads a word beginning with the first letter and reads from left to right until it has reached the last letter. We need our Turing machine to do the same.

$$
\overset{s_0}{\underset{}{}}
$$

$$\# \quad \underline{a_1} \quad a_2 \quad a_3 \quad a_4 \quad a_5 \quad a_6 \quad a_7$$

and we are ready to begin.

We have another way of representing a Turing machine which makes it look more like an automaton. We shall represent the rule

$$(s_i, a, s_j, b, R)$$

by the symbol

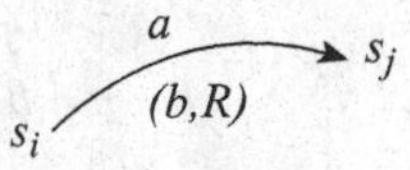

so that the program go-end which has rules

$$(s_0, a, s_1, a, R) \qquad (s_0, b, s_1, b, R) \qquad (s_1, a, s_1, a, R)$$
$$(s_1, b, s_1, b, R) \qquad (s_1, \#, h, \#, \#, \#)$$

may be represented by

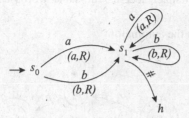

Notice that the letter above the arrow is the letter which is read by the machine. We really do not care what is printed out. We could print a # in each square as it is read or we could simply print back the letter that is read. We shall choose to do the latter. Each time a letter is read, we wish the machine to move one square to the left, so that the next letter is read.

We are now ready to imitate an automaton. If the symbol

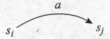

occurs in an automaton, we shall imitate it with the rule

$$(s_i, a, s_j, a, R)$$

or the symbol

It may be recalled that a word is accepted by an automaton if, after the word is read, the automaton is in an acceptance state. For every acceptance state s of the automaton, we will add a rule

$$(s, \#, h, \#, \#)$$

shown as

so that if the word is accepted by the automaton it will also end up in state s of the Turing machine, read the # in front of the word and halt. Thus the Turing machine halting will mean that it accepts a word. Further a Turing machine programmed in this manner accepts the same words as the automaton it is imitating.

For example, given the automaton

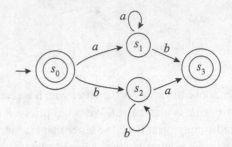

we have the corresponding program for the Turing machine

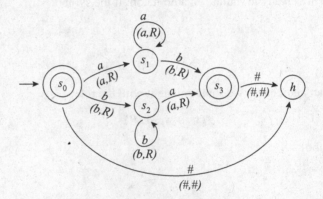

Since a Turing machine can be programmed to accept the same language as a given automaton, we have the following theorem:

Theorem 5.1 *Every regular language is recognized by a Turing machine.*

Definition 5.3 *The languages recognized by Turing machines are called **recursively enumerable**.*

We have already shown that regular languages are recursively enumerable and claimed that context-free languages are recursively enumerable. At this point we shall show how a Turing machine recognizes the language $\{a^n b^n : n$ is a positive integer$\}$, which is context-free, and how it recognizes $\{a^n b^n c^n : n$ is a positive integer$\}$, which is not context-free.

We begin by designing a program for a Turing machine that will recognize the language $\{a^n b^n : n$ is a positive integer$\}$. We basically want the Turing machine to read an a, then read a b, return to read an a, and continue until all of the as and bs have been read, if there is an equal number of them. We begin by reading an a in the first square. We want to know that we have counted this a, so we shall change it to A. We do this with rule

$$(s_0, a, s_1, A, R).$$

We now want to go right until we reach a b, which we shall change to a B. To get to b, we need to pass over each a without changing it and also after the first time we may have to pass over Bs without changing them to reach a b. We do this with the rules

$$(s_1, a, s_1, a, R)$$
$$(s_1, B, s_1, B, R).$$

When we reach a b, we want to change it to a B and go back left. We do that with the rule

$$(s_1, b, s_2, B, L).$$

We now need to go back to find the second a. To do this we go left until we reach an A. This will tell us that the next letter to the right should be the next a. To go back, we need to pass over Bs, and as to get to A. We do this with the rules

$$(s_2, B, s_2, B, L) \qquad (s_2, a, s_2, a, L).$$

When we reach A, we want to go one square to the right to read another a, if there is one. We do this with the rule

$$(s_2, A, s_0, A, R).$$

This puts us back into the cycle of reading another a and another b. If we run out of bs before we run out of as the system will be in state s_1 and eventually try to read a blank so it will hang. If we have read the last a, then when we reach A and go right one square, we will read a B. At this point we need to check to see if there is another b. First we change state if we are in s_0 and read a B. We do this with rule

$$(s_0, B, s_3, B, R).$$

In state s_3, read nothing but Bs and a blank. Thus we have the rules

$$(s_3, B, s_3, B, R) \qquad (s_3, \#, h, \#, \#).$$

This may also be shown as the labeled graph

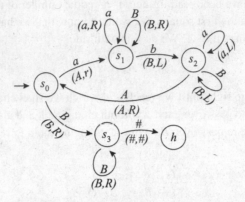

For example consider the string *aabb*. The initial configuration is

$$0$$
$$\underline{a} \quad a \quad b \quad b.$$

We then have the following rules and configurations:

$$(s_0, a, s_1, A, R) \vdash \quad \begin{matrix} & 1 & \\ A & \underline{a} & b & b \end{matrix}$$

$$\Rightarrow (s_1, a, s_1, a, R) \vdash \quad \begin{matrix} & & 1 & \\ A & a & \underline{b} & b \end{matrix}$$

$$\Rightarrow (s_1, b, s_2, B, L) \vdash \quad \begin{matrix} & 2 & & \\ A & \underline{a} & B & b \end{matrix}$$

$$\Rightarrow (s_2, a, s_2, a, L) \vdash \quad \begin{matrix} 2 & & & \\ \underline{A} & a & B & b \end{matrix}$$

$$\Rightarrow (s_2, A, s_0, A, R) \vdash \quad \begin{matrix} & 0 & & \\ A & \underline{a} & B & b \end{matrix}$$

$$\Rightarrow (s_0, a, s_1, A, R) \vdash \quad \begin{matrix} & & 1 & \\ A & A & \underline{B} & b \end{matrix}$$

$$\Rightarrow (s_1, B, s_1, B, R) \vdash \quad \begin{matrix} & & & 1 \\ A & A & B & \underline{b} \end{matrix}$$

$$\Rightarrow (s_1, b, s_2, B, L) \vdash \quad \begin{matrix} & & 2 & \\ A & A & \underline{B} & B \end{matrix}$$

$$\Rightarrow (s_2, B, s_2, B, L) \vdash \quad \begin{matrix} & & 2 & \\ A & A & \underline{B} & B \end{matrix}$$

$$\Rightarrow (s_2, B, s_2, B, L) \vdash \quad \begin{matrix} & 2 & & \\ A & \underline{A} & B & B \end{matrix}$$

$$\Rightarrow (s_2, A, s_0, A, R) \vdash \quad \begin{matrix} & & 0 & \\ A & A & \underline{B} & B \end{matrix}$$

$$\Rightarrow (s_0, B, s_3, B, R) \vdash \quad A \quad A \quad B \quad \underset{\underline{\quad}}{\overset{3}{B}}$$

$$\Rightarrow (s_3, B, s_3, B, R) \vdash \quad A \quad A \quad B \quad B \quad \underset{\underline{\quad}}{\overset{3}{\#}}$$

$$\Rightarrow (s_3, \#, h, \#, \#) \vdash \quad A \quad A \quad B \quad B \quad \underset{\underline{\quad}}{\overset{h}{\#}}.$$

Next we design a program for a Turing machine that will recognize the language $\{a^n b^n : n$ is a positive integer$\}$.

We begin by reading an a in the first square. We want to know that we have counted this a, so we shall change it to a A. We do this with rule

$$(s_0, a, s_1, A, R).$$

We now want to go right until we reach a b, which we shall change to a B. To get to b, we need to pass over each a without changing it and also after the first time we may have to pass over Bs without changing them to reach a b. We do this with the rules

$$(s_1, a, s_1, a, R) \qquad (s_1, B, s_1, B, R).$$

When we reach a b, we want to change it to a B and start back to look for another a. We do this with the rule

$$(s_1, b, s_2, B, L).$$

To go back, we need to pass over Bs and as to get to A. We do this with the rules

$$(s_2, B, s_2, B, L) \qquad (s_2, a, s_2, a, L).$$

When we reach A, we want to go one square to the right to read another a, if there is one. We do this with the rule

$$(s_2, A, s_0, A, R).$$

This puts us back into the cycle of reading another a and b. If we run out of bs before we run out of as the system will hang. If we have read the last a, then when we reach A and go right one square, we will read a B. At this point we need to check to see if there is another b. First we change state if we are in s_0 and read a B. We do this with rule

$$(s_0, B, s_3, B, R).$$

In state s_3, we expect to read nothing but Bs, b, and a blank. Thus we have the rules

$$(s_3, B, s_3, B, R) \qquad (s_3, b, s_4, B, R) \qquad (s_4, \#, h, \#, \#).$$

We now design a program for a Turing machine that will recognize the language $\{a^n b^n c^n : n$ is a positive integer$\}$. In a manner similar to the previous example we want the Turing machine to read an a, then read a b, then read a c, and continue until all of the as, bs, and cs have been read, if there are an equal number of them. We begin by reading an a in the first square. We want to know that we have counted this a, so we shall change it to A. We do this with rule

$$(s_0, a, s_1, A, R).$$

We now want to go right until we reach a b, which we shall change to a B. To get to b, we need to pass over each a without changing it and also after the first time we may have to pass over Bs without changing them to reach a b. We do this with the rules

$$(s_1, a, s_1, a, R) \qquad (s_1, B, s_1, B, R).$$

When we reach a b, we want to change it to a B and continue onward. We do that with the rule

$$(s_1, b, s_2, B, R).$$

We now need to continue until we find a c. We will need to pass over bs and Cs. We do this with the rules

$$(s_2, b, s_2, b, R) \qquad (s_2, C, s_2, C, R).$$

We next want to read c, replace it with a C, and start back to look for another a. We do this with the rule

$$(s_2, c, s_3, C, L).$$

To go back, we need to pass over Cs, bs, Bs, and as to get to A. We do this with the rules

$$(s_3, C, s_3, C, L) \qquad (s_3, b, s_3, b, L) \qquad (s_3, B, s_3, B, L) \qquad (s_3, a, s_3, a, L).$$

When we reach A, we want to go one square to the right to read another a, if there is one. We do this with the rule

$$(s_3, A, s_0, A, R).$$

This puts us back into the cycle of reading another a, b, and c. If we run out of bs or cs before we run out of as the system will hang. If we have read the last a, then when we reach A and go right one square, we will read a B. At this point we need to check to see if there is another b. First we change state if we are in s_0 and read a B. We do this with rule

$$(s_0, B, s_4, B, R).$$

In state s_4, we expect to read nothing but Bs, Cs, and a blank. Thus we have the rules

$$(s_4, B, s_4, B, R) \qquad (s_4, C, s_4, C, R) \qquad (s_4, \#, h, \#, \#).$$

This may also be shown as the labeled directed graph

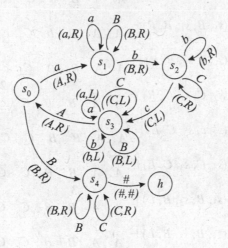

For example consider the string *aabbcc*. The initial configuration is

$$
\begin{array}{ccccccc}
0 & & & & & \\
\underline{a} & a & b & b & c & c
\end{array}
$$

We then have the following rules and configurations:

$$
(s_0, a, s_1, A, R) \vdash \begin{array}{ccccccc} & 1 & & & & \\ A & \underline{a} & b & b & c & c \end{array}
$$

$$
\Rightarrow (s_1, a, s_1, a, R) \vdash \begin{array}{ccccccc} & & 1 & & & \\ A & a & \underline{b} & b & c & c \end{array}
$$

$$
\Rightarrow (s_1, b, s_2, B, R) \vdash \begin{array}{ccccccc} & & & 2 & & \\ A & a & B & \underline{b} & c & c \end{array}
$$

$$
\Rightarrow (s_2, b, s_2, b, R) \vdash \begin{array}{ccccccc} & & & & 2 & \\ A & a & B & b & \underline{c} & c \end{array}
$$

$$
\Rightarrow (s_2, c, s_3, C, L) \vdash \begin{array}{ccccccc} & & & 3 & & \\ A & a & B & \underline{b} & C & c. \end{array}
$$

$$
\Rightarrow (s_3, b, s_3, b, L) \vdash \begin{array}{ccccccc} & & 3 & & & \\ A & a & \underline{B} & b & C & c \end{array}
$$

$$
\Rightarrow (s_3, B, s_3, B, L) \vdash \begin{array}{ccccccc} & 3 & & & & \\ A & \underline{a} & B & b & C & c \end{array}
$$

$$\Rightarrow (s_3, a, s_3, a, L) \vdash \quad \overset{3}{\underline{A}} \quad a \quad B \quad b \quad C \quad c$$

$$\Rightarrow (s_3, A, s_0, A, R) \vdash \quad A \quad \overset{0}{\underline{a}} \quad B \quad b \quad C \quad c$$

$$\Rightarrow (s_0, a, s_1, A, R) \vdash \quad A \quad A \quad \overset{1}{\underline{B}} \quad b \quad C \quad c$$

$$\Rightarrow (s_1, B, s_1, B, R) \vdash \quad A \quad A \quad B \quad \overset{1}{\underline{b}} \quad C \quad c$$

$$\Rightarrow (s_1, b, s_2, B, R) \vdash \quad A \quad A \quad B \quad B \quad \overset{2}{\underline{C}} \quad c$$

$$\Rightarrow (s_2, C, s_2, C, R) \vdash \quad A \quad A \quad B \quad B \quad C \quad \overset{2}{\underline{c}}$$

$$\Rightarrow (s_2, c, s_3, C, L) \vdash \quad A \quad A \quad B \quad B \quad \overset{3}{\underline{C}} \quad C$$

$$\Rightarrow (s_3, C, s_3, C, L) \vdash \quad A \quad A \quad B \quad \overset{3}{\underline{B}} \quad C \quad C$$

$$\Rightarrow (s_3, B, s_3, B, L) \vdash \quad A \quad A \quad \overset{3}{\underline{B}} \quad B \quad C \quad C$$

$$\Rightarrow (s_3, B, s_3, B, L) \vdash \quad A \quad \overset{3}{\underline{A}} \quad B \quad B \quad C \quad C$$

$$\Rightarrow (s_3, A, s_0, A, R) \vdash \quad A \quad A \quad \overset{0}{\underline{B}} \quad B \quad C \quad C$$

$$\Rightarrow (s_0, B, s_4, B, R) \vdash \quad A \quad A \quad B \quad \overset{4}{\underline{B}} \quad C \quad C$$

$$\Rightarrow (s_4, B, s_4, B, R) \vdash \quad A \quad A \quad B \quad B \quad \overset{4}{\underline{C}} \quad C$$

$$\Rightarrow (s_4, C, s_4, C, R) \vdash \quad A \quad A \quad B \quad B \quad C \quad \overset{4}{\underline{C}}$$

$$\Rightarrow (s_4, C, s_4, C, R) \vdash \quad A \quad A \quad B \quad B \quad C \quad C \quad \overset{4}{\underline{\#}}$$

$$\Rightarrow (s_4, \#, h, \#, \#) \vdash \quad A \quad A \quad B \quad B \quad C \quad C \quad \overset{h}{\underline{\#}}.$$

We now show how to perform two arithmetic operations on a Turing machine. The first of these is addition, which is trivial. Suppose we have p, represented by a string of 1s of length p, and q represented by a string of 1s of length q, so we have the configuration

$$\overset{0}{1} \quad 1 \quad 1 \quad \dots \quad 1 \quad \# \quad 1 \quad 1 \quad 1 \quad 1 \quad \dots \quad 1.$$

We simply use *delete*(#) to delete the blank, and we have

0

1 1 1 ... 1 1 1 1 1 ... 1

which is a string of 1s of length $p + q$, which represents $p + q$.

We now sketch a method of multiplying positive integers p and q, which are represented by strings of 1s of lengths p and q respectively. Details of the multiplication are left to the reader. We begin with the configuration

0

1 1 1 ... 1 # 1 1 1 1 ... 1.

Replace the first 1 with β. Replace the second 1 with β (if there is no second 1, move left and delete β, leaving only the string q). Otherwise, move to the end of the string of 1s of length q. Place a blank (so that the length of q is retained), and place another string of 1s of length q after the blank. Return to β and then go right to the next 1 in the string for p. Replace the 1 with β and again place a string of 1s of length q at the end of the third string. Continue this until there are no more 1s in the string for p. At that point when the machine tries to read a 1 from p, it will read a #. Go to end of the third string. Go left deleting all blanks. Then continue left until reaching a β. Delete all βs to produce the answer.

Exercises

(1) Supply the details for the Turing machine program *delete*(c).
(2) Design a Turing machine for the program *go-left*(n) which moves the head of the machine to the left n squares.
(3) Design a Turing machine for *insert*(c) which begins at the square of the letter we were replacing and returns to that square. (Hint: Instead of placing the c in the square, we would place a marker, and then when we had finished moving the letters we would return to the marker and replace it with a c.)
(4) Design a Turing machine for *delete*(c) where the machine moves over to the square containing c and replaces it with a marker which is not part of the regular alphabet. It then goes to the end of the string and moves each letter to the left in a similar manner to the one we use to move letters to the right in *insert*(c). It replaces the marker with the letter to its right and then changes states to remain where the letter was inserted.
(5) Design a Turing machine that multiplies two positive integers.
(6) Design a Turing machine that subtracts a smaller number from a larger one.
(7) Design a Turing machine that accepts the language described by the expression $ab^*c^*(b \vee ac)$.

(8) Design a Turing machine that accepts the language described by the expression $abc(b \lor ac)^*b$

(9) Design a Turing machine that accepts all strings in a and b except aba and abb.

(10) Design a Turing machine that accepts the language described by the expression $(aa^*bb^*)^*$.

(11) Write the list of rules and configurations to describe the results when the program that accepts $a^n b^n$ for a positive number n tries to read $a^2 b^3$.

(12) Write the list of rules and configurations to describe the results when the program that accepts $a^n b^n$ for a positive number n tries to read $a^3 b^2$.

(13) Write the list of rules and configurations to describe the results when the program which accepts $a^n b^n c^n$ for a positive number n tries to read $a^3 b^2 c^3$.

(14) Design a Turing machine that accepts the language of all strings in a and b that have the same number of as and bs.

(15) For a given string s consisting of as and bs, define **reverse**(s) to be the string s written backwards. Thus $reverse(abbb) = bbba$. Design a Turing machine that, given a string s, prints its reverse.

(16) A **palindrome** over the set $\{a, b, c\}$ is a string such that $s = reverse(s)$. Thus $abbcbba, abba, abcba$, and $cbaabc$ are palindromes. An **even palindrome** has an even number of letters in the string and an **odd palindrome** has an odd number of letters in the string. Design a Turing machine that accepts all even palindromes.

(17) Design a Turing machine that accepts all odd palindromes.

(18) Design a Turing machine that accepts all words of the form $a^n b^n a^n$ for any positive integer n.

(19) Design a Turing machine that accepts all words of the form ww where w is a string of as and bs.

5.2 Nondeterministic Turing machines and acceptance of context-free languages

We begin by showing that a context-free language can be accepted by a nondeterministic Turing machine and then show that any language accepted by a nondeterministic Turing machine is accepted by a deterministic Turing machine.

Definition 5.4 *A Turing machine is not deterministic if δ is a finite subset of* $(Q \times \Gamma) \times (Q \times \Gamma \times N)$.

Thus δ is replaced by a relation, which we shall denote by θ. Thus $\theta(s, a)$ is a subset of $(Q \times \Gamma \times N)$. Since a context-free language is accepted by a pushdown automata, we show that a pushdown automaton can be imitated by a Turing machine and hence this Turing machine accepts a context-free language. The only problem at this point is that a pushdown automaton is not deterministic and hence the Turing machine we create is not deterministic. Thus we must show that any language accepted by a nondeterministic Turing machine is accepted by a deterministic Turing machine.

Theorem 5.2 *A context-free language is accepted by a nondeterministic Turing machine.*

Proof We shall prove this theorem informally assuming that the steps in the conversion can be easily replaced by subroutines in a Turing machine. When at a given step, we have the word w to read and ω in the stack of the pushdown automata, we shall associate this with $w \nabla \omega$ on the tape of the Turing machine. The word w is accepted if $w \nabla \#$ is converted to $\# \nabla \#$. Assume the tape begins with a blank followed by the word. Assume also that the Turing machine is positioned at the first letter of the word.

For each of the rules for a pushdown automaton, we shall give the corresponding instructions for a Turing machine.

1	$((a, s, E), (t, D))$	In state s, a is read and E is popped, go to state t and push D.
2	$((a, s, \lambda), (t, D))$	In state s, a is read, go to state t and push D.
3	$((\lambda, s, \lambda), (s, D))$	In state s, push D.
4	$((a, s, E), (t, \lambda))$	In state s, and a is read, pop E and go to state t.
5	$((\lambda, s, E), (s, \lambda))$	In state s, pop E.
6	$((a, s, \lambda), (t, \lambda))$	In state s, read a and go to state t.
7	$((a, s, \lambda), (s, \lambda))$	In state s, read a.

1. Go to the first position after ∇, delete E and insert D. Return to a and delete a. Go to state t.
2. Go to the first position after ∇, insert D. Return to a and delete a. Go to state t.
3. Go to the first position after ∇, insert D. Return to the original letter. Go to state s.
4. Go to the first position after ∇, delete E. Return to a and delete a. Go to state t.
5. Go to the first position after ∇, delete E. Return to the original position. Go to state s.
6. Delete a. Go to state t.
7. Delete a. Go to state s.

□

We now show that a language accepted by a nondeterministic Turing machine T is accepted by a deterministic Turing machine T'. Since T is nondeterministic, and, $\theta(s, x)$ is a subset of $(Q \times \Gamma \times N)$, if $\theta(s, x)$ contains k elements, we shall denote them by $\theta_0(s, x), \theta_1(s, x), \theta_2(s, x) \ldots \theta_{k-1}(s, x)$. Thus each of the $\theta_i(s, x)$ is well defined. For example we could have $\theta_0(s, x) = (s', b, R)$, so the rule is (s, x, s', b, R) and $\theta_1(s, x) = (s'', C, L)$, so the rule is (s, x, s'', C, L). We number the elements in each $\theta(s_i, a_i)$ for each state s_i and each a_i in Σ. Hence if we are in state s and and have input x, and are given an integer j, we can use $\theta_j(s, x)$ to supply the rule to use. Assume that we never need more than $n + 1$ integers to label the subsets for any $\theta(s_i, a_i)$, then if we have a sequence of nonnegative integers $m_1, m_2, \ldots, m_p$ less than or equal to n, we could sequentially apply $\theta_{m_1}, \theta_{m_2}, \ldots, \theta_{m_p}$, which together with the state and input would give us the rules to use. If we apply all possible relevant sequences, we can produce all possible computations. Hence if a word is accepted by the Turing machine T, it will be accepted in one of these computations.

The next problem is the production of the sequences of integers. We shall label these sequences $N_0, N_1, N_2, \ldots, N_i, \ldots$ We begin with $N_0 = 0$ and simply count in base $n + 1$. Thus the sequences are

$$(0), (1), (2), \ldots, (n), (1, 0), (1, 1), (1, 2), (1, 3), \ldots, (1, n), (2, 0), \ldots$$

The sequence following

$$(1, 3, 4, 3, 2, 3)$$

is

$$(1, 3, 4, 3, 2, 4),$$

and the sequence following

$$(1, 3, 4, n, n, n)$$

is

$$(1, 3, 5, 0, 0, 0).$$

The subroutine in which a Turing machine changes the number N_k to N_{k+1} is straightforward and is left to the reader in the problems, with the warning that as the length of the sequence is increased, it is increased to the right on the tape.

We next have to decide how to proceed in reading a given word. We will place the word to be read, followed by | and the current sequence on the tape. At each step we will mark the letter being read, keeping track of the state the machine is in, and then proceed to the right to locate the proper number in the sequence. If the machine is in state s, and we mark a, we shall use a_s as the marker. Thus we retain information about both the letter read and the current state of the machine. As we use a number in the sequence we mark it with a $'$, so that we proceed each time to the first unmarked number in the sequence, select it, mark it, and then return to the marked letter with the information needed to select the proper path for the Turing machine to take, given the state and the letter being read. For example suppose the Turing machine is in state s, reads letter a, and finds that j is the number selected, it then proceeds with $\theta_j(s, a)$ to supply the rule to use.

As an illustration suppose we have

$$a_{s_1} \quad b_{s_2} \quad a_{s_3} \quad \overset{s_4}{\underline{b}} \quad b \quad a \quad a \quad | \quad 1' \quad 2' \quad 2' \quad 1 \quad 3 \quad 1 \quad 2.$$

We change b to b_{s_4} so we have

$$a_{s_1} \quad b_{s_2} \quad a_{s_3} \quad \overset{s_4}{\underline{b_{s_4}}} \quad b \quad a \quad a \quad | \quad 1' \quad 2' \quad 2' \quad 1 \quad 3 \quad 1 \quad 2.$$

We then move to 1, the first unmarked integer and mark it, so we have

$$a_{s_1} \quad b_{s_2} \quad a_{s_3} \quad \overset{s_4}{b_{s_4}} \quad b \quad a \quad a \quad | \quad 1' \quad 2' \quad 2' \quad \overset{q_1}{1'} \quad 3 \quad 1 \quad 2$$

where the subscript of the state is the number selected. We then return to b_{s_4} where we have $\theta^*(q_1, b_{s_4}) = \theta_1(s_4, b)$.

The instructions could be as follows

$\theta^*(s_i, \alpha) = \theta_{s_i}(t_1, \alpha_{s_i}, R)$

$\theta^*(t_1, x) = (t_1, x, R)$ for $x \neq |$

$\theta^*(t_1, |) = (t_2, |, R)$

$\theta^*(t_2, n') = (t_2, n', R)$ for integer n'

$\theta^*(t_2, m) = (q_m, m', L)$ for first unmarked integer m

$\theta^*(q_m, n') = (q_m, n', L)$ for all marked integers n'

$\theta^*(q_m, x) = (q_m, x, L)$ if x is an unmarked letter of the alphabet

$\theta^*(q_m, \alpha) = \theta_m(s_i, \alpha)$ if a is marked by s_i.

Informally we state the procedure for testing a word for acceptance by a Turing machine as follows: First, given the word, duplicate the word and follow it by the first sequence so that we have

$$w\#w \mid 0.$$

Perform the process above for testing the second copy of the word w following the sequence #. At the beginning, the machine is positioned at the first letter of w. If $\theta^*(t_2, \#)$ occurs and the word is not accepted, the end of the sequence has been reached. Erase the symbols between # and |, again duplicate w after the #, proceed to the next sequence and repeat the process until the word is accepted or the length of the sequence exceeds m^{n+1}, where there are m productions beginning with $\theta^*(s_i, \alpha_i)$ and n is larger than the number of productions beginning with any $\theta^*(s_i, \alpha_i)$ in the nondeterministic Turing machine of the word, since all possibilities have been tried. Since the new Turing machine which we shall call T' just defined is deterministic and a word is accepted by T' if and only if it is accepted by T, we have shown that a word is accepted by a nondeterministic Turing machine if and only if it is accepted by a deterministic Turing machine.

We finally conclude that if a word is context-free, it is accepted by a Turing machine.

Exercises

(1) In Theorem 5.2 write a subroutine for producing the sequence of integers used for showing that a language accepted by a nondeterministic Turing machine can be accepted by a deterministic Turing machine.

Find Turing machines (not necessarily deterministic) that accept the context-free languages.
(2) The language containing twice as many as as bs.
(3) The language containing the same number of as and bs.
(4) The language $\{a^n b^n : n = 1, 2, \dots\}$.
(5) The language $\{a^n b^k c^n : k, n = 1, 2, \dots\}$.
(6) The language of palindromes of odd length on the alphabet $\{a, b, c\}$.
(7) The language of palindromes of even length on the alphabet $\{a, b, c\}$.
(8) The language of all palindromes on the alphabet $\{a, b, c\}$.
(9) The language $\{a^n b^n a^m b^m : m, n = 1, 2, \dots\}$.
(10) The language $\{a^n b^{2n} a^m b^{2m} : m, n = 1, 2, \dots\}$.

5.3   The halting problem for Turing machines

One of the more frustrating problems running a computer problem occurs when the computer continues to run with no end in sight. One has the dilemma of deciding whether the computer has just not finished the problem, and perhaps in five minutes or five hours it will finish the problem, or if it is in a loop and will continue to run forever. This would be particularly true if the machine were as inefficient as a Turing machine. It would be nice if one could determine in advance whether the machine was going to halt, which is equivalent to solving the problem. Assuming Church's Thesis, if there was an algorithmic step by step way of determining whether a machine was going to halt, then a program could be written for a Turing machine that could determine whether a machine was going to halt.

This particular problem is called the **halting problem** and is formally stated as follows:

Halting Problem Is there an algorithm which will determine whether, for any given Turing machine T and any input string w, the Turing machine T, given the input string w, will reach the halt state?

Before answering this question, we look at some related properties of a Turing machine. When we were looking at acceptance of regular languages by a Turing machine, a word was accepted if the machine reached the halt state. Otherwise, the machine crashes, hangs, or loops. Any language which is accepted in this manner is called **Turing acceptable**.

It would be nice if the Turing machine, when a word was read by it, would print Y at the beginning of the tape if the word were in the language and N if the word were not in the language.

Definition 5.5 *A language L is **Turing decidable** if there exists a Turing machine that, when a string is input, prints Y on the tape if the word is in L and N if the word is not in L.*

Theorem 5.3 *If a language is Turing decidable then it is Turing acceptable.*

Proof If a language L is Turing decidable, then there is a Turing machine that prints Y if the word is in L and N if the word is not in L. Modify this machine so that instead of printing N, it goes into an infinite loop and instead of printing Y, goes into the halt state. Thus the new machine halts if a word is in L and goes into an infinite loop if the word is not in L. Thus L is Turing acceptable.
□

Theorem 5.4 *If a language L is Turing decidable, then its complement $L' = A^* - L$ is Turing decidable.*

Proof If a language L is Turing decidable, then there is a Turing machine that prints Y if the word is in L and N if the word is not in L. Modify this machine so that instead of printing Y, it prints N and instead of printing N, it prints Y. This new machine prints Y if the word is in L' and N if the word is not in L'. Thus L' is Turing decidable.
□

Theorem 5.5 *A language L is Turing decidable if and only if both L and L'*
are Turing acceptable.

Proof If a language L is Turing decidable then by Theorem 5.4, its complement L' is also Turing decidable. But by Theorem 5.3, L and L' are then both
Turing acceptable.

Conversely, if L and L' are both Turing acceptable, then there are machines
M and M' that accept languages L and L' respectively. Place the input string
in both M and M'. If the input string is accepted by M, then print the letter Y.
If the input string is accepted by M', then print the letter N. Since this process
is algorithmic, by Church's Thesis, it can be duplicated by a Turing machine
M''. Hence L is Turing decidable and, by Theorem 5.4, its complement L' is
also Turing decidable.   □

Before proceeding further we need to show that every Turing machine with
alphabet $A = \{a, b\}$ can be uniquely described by a string of as and bs. It is
obvious that a Turing machine is uniquely determined by the set of rules for
the machine. We shall show this for the set of states $S = \{s_1, s_2, s_3, \ldots, s_n\}$. It
may be recalled that a rule has the form

$$(s_i, a, s_j, b, L)$$

where the first and third components are states, the second and fourth components are letters of a set of tape symbols Γ which contains the alphabet. We may
also have #, and Δ, which may be used as a marker in Γ. The last component is
either L or R. At times we have also included # in the last component, but this
was not really necessary. It was merely used in the halt statement to indicate
that the machine had halted and so had ceased moving. We proceed with the
encoding as follows: If the first component is s_i, we begin with a string of as
of length i. Thus if the first component is s_3, we begin with aaa. We follow
this with a b, which is used as a divider. For the symbols a, b, #, and Δ, we
add to the string aa, bb, ab, and ba respectively. Thus if the first component is
s_4 and the next component is a, then our string at this point is $aaaabaa$. We
follow this with the string of as corresponding to the state s_j and then another
b for a divider. We include a for L and b for R as the fourth component. Thus
the string for $(s_4, b, s_2, \#, R)$ is $aaaabbbaababb$, where $aaaa$ represents s_4, b
is a divider, bb represents b, aa represents s_2, b is a divider, ab represents #,
and b represents R. Once we have a string for each rule, we then concatenate
or connect all of the strings together to form one long string of as and bs that
represent the Turing machine.

It is also possible to decode the string. For example suppose we had the string *ababaaabbbb* . . . , the first *a* represents s_1, the *b* is a divider, *ab* represents #, *aaa* represents s_3, *b* is a divider, *bb* represents *b*, and *b* represents *R*. We have decoded the rule $(s_1, \#, s_3, b, R)$ and we continue reading the string to get the next rule. Denote the string that represents the Turing machine *M* by $c(M)$.

Suppose we want to display a string that represents a Turing machine followed by input to be read by the machine. This is easily done by taking the string for the Turing machine *M* followed by a *b* and then followed by the input. Since no rule starts with a *b*, finding a *b* would indicate to the decoder that data followed rather than another rule. Thus the string representing machine *M* with input *w* is $c(M)bw$.

Consider the language L_0 which consists of all strings $c(M)bw$ representing a Turing machine *M* followed by an input word *w* where the Turing machine accepts that input word. It is simple to construct a Turing machine *MM* that accepts L_0. Given a string *t*, *MM* first decodes *t* and if it represents a Turing machine *M* followed by input data, it inputs the data into the machine *M*, which can be recovered from the string *t*, and *MM* accepts the input $c(M)bw$ if and only if *M* accepts the input string *t*. Therefore L_0 is Turing acceptable. If, in addition L_0 is Turing decidable, then every Turing acceptable language is Turing decidable. To show this we know that if L_0 is Turing decidable, then there exists a Turing machine, say MM_2, which, given any input string *w*, will print *Y* if *w* is in L_0 and *N* if *w* is not in L_0. Assume that we have a Turing acceptable language *L*, then it is accepted by a Turing machine $M(L)$. We can now construct a Turing machine $M'(L)$ which, given an input string *s*, prints *Y* if *s* is in *L* and *N* if *s* is not in *L*. The Turing machine $M'(L)$ is constructed by simply taking the string $c(M(L))$, adding the input string *s* to form $c(M(L))bs$ and then using it as an input string *s'* for MM_2. If MM_2 prints *Y* for the input *s'* then machine $M(L)$ accepts *s*, so $M'(L)$ prints *Y*. If MM_2 prints *N* for the input *s'* then machine $M(L)$ does not accept *s*, so $M'(L)$ prints *N*. Thus *L* is Turing decidable. It thus follows that every acceptable language is decidable if and only if L_0 is decidable.

We now show that L_0 is not Turing decidable. We claim that if L_0 is Turing decidable then the language $L_1 = \{c(M)$ such that *M* accepts $c(M)$ is Turing decidable$\}$. To show this, assume that L_0 is Turing decidable. We construct M_1 as follows: Given an input string *s*, we simply take the string *sbs* and use it as input for MM_2. If MM_2 prints *Y* then $s = c(M)$ for some *M* that accepts $c(M)$. Therefore $s \in L_1$ and M_1 prints *Y*. If MM_2 prints *N* then $s \neq c(M)$ for any machine *M* that accepts $c(M)$, so that $s \notin L_1$ and M_1 prints *N*. Thus L_1 is Turing decidable.

We now show that L_1 is not Turing decidable. Since by Theorem 5.4, L_1 is Turing decidable if and only if L'_1 is, we shall prove that L'_1 is not Turing decidable. The language $L'_1 = \{\{w : w \in \{a, b\}^*\}$ and either $w \neq c(M)$ for any machine M or $w = c(M)$ for some machine M but M does not accept $w\}$. Here we are reminded of Russell's paradox. Let M'_1 be a machine that accepts L'_1, then is it true that $c(M'_1) \in L'_1$? If so then M'_1 does not accept $c(M'_1)$ by definition of L'_1. But M'_1 does accept $c(M'_1)$ because it accepts every element of L'_1, and we have a contradiction. Conversely assume $c(M'_1) \notin L'_1$. Then $c(M'_1) \in L_1$ so that M'_1 accepts $c(M'_1)$ by definition of L_1. But M'_1 only accepts elements of L'_1 so that $c(M'_1) \in L'_1$, again a contradiction. Hence L'_1 is not Turing acceptable and certainly not Turing decidable.

Since we have shown that L_0 is Turing acceptable but not Turing decidable, we have the following theorem:

Theorem 5.6 *There exists a language that is Turing acceptable but not Turing decidable.*

Since L_0 is Turing acceptable but not Turing decidable, the following theorem follows from Theorem 5.5:

Theorem 5.7 *There exists a language which is Turing acceptable but whose complement is not Turing acceptable.*

We have also solved the halting problem since a string is acceptable if and only if the machine reaches the halt state. Hence the algorithm that would satisfy the halting problem is the algorithm which describes MM_2 and it does not exist.

Theorem 5.8 *Given a Turing machine T and an input string w, there is no algorithm which will determine whether the Turing machine T, given the input string w, will reach the halt state.*

Exercises

(1) Show that a finite set is Turing decidable.
(2) Find the string representing the rule (s_5, Δ, s_2, a, R).
(3) Find $c(M)$ where M is the machine defined by the rules

$$(s_1, a, s_2, a, R) \qquad (s_1, b, s_2, b, R) \qquad (s_2, a, s_2, a, R)$$
$$(s_2, b, s_2, b, R) \qquad (s_1, \#, s_3, \#, R).$$

(4) Find $c(M)$ where M is the machine defined by the rules

$$(s_1, a, s_2, \#, R) \qquad (s_1, b, s_2, a, R) \qquad (s_2, a, s_2, \#, R)$$
$$(s_2, b, s_2, a, R) \qquad (s_1, \#, s_3, \#, R).$$

(5) Find the rule that corresponds to the string *aaabababbaa*.

(6) Find the rule that corresponds to the string *aabbbaaabbab*.

(7) Which of the following strings correspond to rules?
  (a) *baaabbaabb*
  (b) *aabbbaaabbbb*
  (c) *aababaabbaa*
  (d) *aabaabaabaab*
  (e) *aabaaabbabb*.

(8) Find the Turing machine that corresponds to the string

$$abaaabbbbaabbbabaababababbaababb.$$

(9) Find the Turing machine that corresponds to the string

$$abaaaababbabbbaababbaababaaababb.$$

(10) Find the Turing machine and input that correspond to the string

$$abaaaabbbbabbbaabbbbaabbbabbbaababaaababbbaaabbb.$$

(11) Find the Turing machine and input that correspond to the string

$$abaaaabbbbabbbaabaabaabbbabbbaaabaaabbbaaababaa$$
$$abababaabb.$$

(12) Devise a method of coding that allows the use of A and B as well as a and b by allowing strings of length 3 to represent input and output symbols.

(13) Use the coding in the previous problem to find the string corresponding to (s_1, a, s_3, A, R).

(14) Find the string that represents the machine

$$(s_1, a, s_2, b, R) \qquad (s_1, b, s_2, b, R) \qquad (s_2, a, s_2, \#, R)$$
$$(s_2, b, s_2, b, R) \qquad (s_1, \#, s_3, \#, R)$$

together with input *ababaab*.

(15) Find the string that represents the machine

$$(s_1, a, s_2, b, R) \qquad (s_1, b, s_2, a, R) \qquad (s_2, a, s_2, \#, R)$$
$$(s_2, b, s_2, \#, R) \qquad (s_1, \#, s_3, \#, R)$$

together with input *babbab*.

(16) Let L be a language. Prove that one and only one of the following must
be true:

    (a) Neither L nor L' is Turing acceptable.

    (b) Both L and L' are Turing decidable.

    (c) Either L or L' is Turing acceptable but not Turing decidable.

5.4   Undecidability problems for context-free languages

We begin with the Post's Correspondence Problem, which is not only an interesting problem in itself, but is used to prove that certain statements about context-free languages are undecidable.

Definition 5.6 *Given an alphabet* Σ, *let P be a finite collection of ordered pairs of nonempty strings* $(u_1, v_1), (u_2, v_2), \ldots, (u_m, v_m)$ *of* Σ. *Thus P is a finite subset of* $\Sigma^+ \times \Sigma^+$. *A* **match** *of P is a string w for which there exists a sequence of pairs* $(u_{i_1}, v_{i_1}), (u_{i_2}, v_{i_2}), \ldots, (u_{i_m}, v_{i_m})$, *such that* $w = u_{i_1} u_{i_2} \cdots u_{i_m} = v_{i_1} v_{i_2} \cdots v_{i_m}$. **Post's Correspondence Problem** *is to determine if a match exists.*

    An alternative way to think about Post's Correspondence Problem is to consider two lists $A = u_1, u_2, \ldots, u_n$ *and* $B = v_1, v_2, \ldots, v_n$ *where each u_i and v_i is a nonempty string of* Σ *and there is a match if there exists w such that* $w = u_{i_1} u_{i_2.} \cdots u_{i_m} = v_{i_1} v_{i_2} \ldots v_{i_m}$. *The important factor is that the products must consist of corresponding pairs.*

Example 5.1 Let $P = \{(a, ab), (bc, cd), (de, ed), (df, f)\}$, then *abcdedf* and *abcdededf* are both matches of P.

    We wish to show that Post's Correspondence Problem is not decidable. To help us do so we define a modified correspondence system. We shall show that if the modified correspondence system is not decidable, then Post's Correspondence Problem is not decidable. Finally we show that the modified correspondence is not decidable.

Definition 5.7 *Given an alphabet* Σ, *Let P be a finite collection of ordered pairs of nonempty strings* $(u_1, v_1), (u_2, v_2), \ldots, (u_m, v_m)$ *of* Σ *together with a special pair* (u_0, v_0). *In a* **modified correspondence system**, *a* **match** *of P is a string w such that there exist a sequence of pairs* $(u_0, v_0), (u_{i_1}, v_{i_1}), (u_{i_2}, v_{i_2}), \ldots, (u_{i_m}, v_{i_m})$, *such that* $w = u_0 u_{i_1} u_{i_2} \cdots u_{i_m} = v_0 v_{i_1} v_{i_2} \ldots v_{i_m}$. *Thus a match must begin with the designated pair* (u_0, v_0). *The* **modified Post's Correspondence Problem** *is to determine if a match exists in a modified correspondence system.*

Note that in the previous example, any match w must begin with (a, ab), however P is not a modified correspondence, since we are not required to begin with (a, ab) to try to form a match.

Lemma 5.1 *If Post's Correspondence Problem is decidable, then the modified Post's Correspondence Problem is decidable.*

Proof Let P_1 be a modified correspondence system with the sequence of ordered pairs $(u_0, v_0)(u_1, v_1), (u_2, v_2), \ldots, (u_m, v_m)$ and the alphabet Σ consist of all symbols occurring in any of u_i or v_i. Assume every match must begin with u_0 and v_0. Assume also that $\star$ and $\$$ do not occur in Σ. For a string $w = a_1 a_2 a_3 \cdots a_k$, define $L(w) = \star a_1 \star a_2 \star a_3 \cdots \star a_k$ and $R(w) = a_1 \star a_2 \star a_3 \star \cdots a_k \star$. Let P_2 contain the pair $(L(u_0), L(v_0)\star)$, and for all other (u, v) in P_1, let $(L(u), R(v))$ belong to P_2. In addition include, $(\star\$, \$)$ in P_2. It is obvious that only $(L(u_0), L(u_0)\star)$ can begin a match in P_2, since it is the only pair where we do not have one word in the pair beginning with a star while the other does not. It is also obvious that the only pair that can end a pair in P_2, is $(\star\$, \$)$, since it is the only word where the last symbols match, that is we do not have one ending in a star while the other does not.

It is also obvious that if there exist a sequence of pairs

$$(u_0, v_0)(u_{i_1}, v_{i_1}), (u_{i_2}, v_{i_2}), \ldots, (u_{i_m}, v_{i_m})$$

in P_1, such that $w = u_0 u_{i_1} u_{i_2} \cdots u_{i_m} = v_0 v_{i_1} v_{i_2} \cdots v_{i_m}$. Then the sequence

$$(L(u_0), L(v_0)\star)(L(u_{i_1}), R(v_{i_1})), (L(u_{i_2}), R(v_{i_2})), \ldots, (L(u_{i_m}), R(v_{i_m})), (\star\$, \$)$$

produces a match

$$w' = L(u_0)L(u_{i_1})L(u_{i_2})\ldots L(u_{i_m}) \star \$ = L(u_0) \star R(v_{i_1})R(v_{i_2})\ldots \$$$

in P_2. The words

$$L(u_0)L(u_{i_1})L(u_{i_2})\ldots L(u_{i_m}) \star \$$$

and

$$L(v_0) \star R(v_{i_1})R(v_{i_2})\ldots \$$$

in P_2 differ from the words $u_0 u_{i_1} u_{i_2} \cdots u_{i_m}$ and $v_0 v_{i_1} v_{i_2} \cdots v_{i_m}$ respectively in P_1 in the fact that that they have stars between the letters and end in $\$$.

Hence, since a match in the modified Post's correspondence system has a corresponding match in Post's correspondence system, if Post's Correspondence Problem is decidable, then the modified Post's Correspondence Problem is decidable. $\square$

Example 5.2 Using the previous modified Post's correspondence

$$P_1 = \{(a, ab), (bc, cd), (de, ed), (df, f)\}$$

with match $abcdedf$, we have

$$P_2 = \{(*a, *a*b*), (*b*c, c*d*), (*d*e, e*d*), (*d*f, f*), (\star\$, \$)\}$$

with match $*a*b*c*d*e*d*f \star \$$.

Theorem 5.9 *Post's Correspondence Problem is undecidable.*

Proof We show that Post's Correspondence Problem is undecidable by showing that the modified Post's Correspondence Problem is undecidable. We do this by showing that if the modified Post's Correspondence Problem is decidable, then L_0 (see previous section) is acceptable, which means that it is decidable if a Turing machine accepts a given word. Assuming the sequence for a given Turing machine and word, we construct a modified Post's correspondence system that has a match if and only if M accepts w. Intuitively assume

$$\#s_0 w \# \alpha_1 s_1 \beta_1 \# \alpha_2 s_2 \beta_2 \# \ldots \# \alpha_k s_k \beta_k \#$$

describes the process used by the Turing machine to read w, where each w, and each of the α_i and β_i, are strings, the Turing machine begins in state s_0. Each of the following steps describes the process for the machine accepting w. Hence between the spaces, each string in the match represents symbols on the tape and the state of the machine for each step as the Turing machine progresses in its computation. We wish to create a modified post's correspondence system which has this description. We shall see that the overlapping produced by the rules below together with the fact that the top and bottom row must match give us the process described above.

Note that reaching an acceptance state is equivalent to reaching a halt state since as above, we can create rules that take us from an acceptance state to the final state.

For a given Turing machine and word, we create the modified Post's correspondence system as follows. We shall use two rows to represent the first coordinates and second coordinates respectively. The following are the rules we are allowed to use. We begin with the pair $(\#, \#s_0 w)$ so that we have

#
#$s_0 w$

Since this is a modified Post's correspondence system, we can require that we begin with this pair. For each X in Γ we have

$$\frac{X}{X}.$$

We next use the following pairs to guide us in selecting the next string in our match:

For each state s, which is not a final state and each state s', and symbols X, Y, and Z in Γ,

$$\frac{sX}{Ys'} \qquad \text{if } \delta(s, X) = (s', Y, R)$$

$$\frac{XsY}{s'XZ)} \qquad \text{if } \delta(s, Y) = (s', Z, L)$$

$$\frac{s\#}{Xs'\#} \qquad \text{if } \delta(s, \#) = (s', X, R)$$

$$\frac{Xs\#}{s'XY\#} \qquad \text{if } \delta(s, \#) = (s', Y, L).$$

We shall call these the pairs generated by δ.

In trying to get our match this set guides us to the next string. For example if we have

$$\ldots\#$$
$$\ldots\#11s_i011\#$$

in our match and one of the pairs above is $(s_i0, 1s_j)$, we will want the next string to be $\#111s_j11\#$. Note however that the two 1s at the beginning and end of the string are not affected by the pair above. Hence we need pairs $(\#, \#)$ and $(1, 1)$ to get

$$\ldots\#11s_i011\#$$
$$\ldots\#11s_i011\#111s_j11\#.$$

More precisely we would use $\dfrac{1}{1}, \dfrac{1}{1}, \dfrac{s_i0}{1s_j}, \dfrac{1}{1}, \dfrac{1}{1}, \dfrac{\#}{\#}$. Hence we need pairs

$$\frac{X}{X} \qquad \text{for all } X \text{ in } \Gamma$$

$$\frac{\#}{\#}.$$

Obviously if we never get to an acceptance state (and hence a final state) we will never have a match since there will always be an overlap at the bottom. We thus need rules to get a match if we reach a halt state h. We use the following pairs to get rid of the overlap.

$$(0s_m0, s_m)$$
$$(1, 1)$$
$$(\#, \#)$$
$$(0s_m1, s_m)$$
$$(1s_m0, s_m)$$
$$(1s_m1, s_m)$$
$$(0s_m, s_m)$$
$$(s_m0, s_m)$$
$$(1s_m, s_m)$$
$$(s_m1, s_m)$$
$$(s_m\#\#, \#).$$

The last term gets rid of the overlap s_m when all of the other symbols have been eliminated. Thus if we reached

$$\ldots\#11s_i011\#$$
$$\ldots\#11s_i011\#111s_m11\#$$

rules $\dfrac{1s_m1}{s_m}$, $\dfrac{1s_m}{s_m}$, $\dfrac{1}{1}$, $\dfrac{\#}{\#}$, and $\dfrac{s_m\#\#}{\#}$ would produce

$$\ldots\#11s_i011\#111s_m11\#11s_m1\#1s_m\#s_m\#\#$$
$$\ldots\#11s_i011\#111s_m11\#11s_m1\#1s_m\#s_m\#\#$$

as follows

$$\ldots\#11s_i011\#$$
$$\ldots\#11s_i011\#111s_m11\#$$

$$\ldots\#11s_i011\#1$$
$$\ldots\#11s_i011\#111s_m11\#1$$

$$\ldots\#11s_i011\#11$$
$$\ldots\#11s_i011\#111s_m11\#11$$

$$\ldots\#11s_i011\#111s_m1$$
$$\ldots\#11s_i011\#111s_m11\#11s_m$$

$$\ldots\#11s_i011\#111s_m11$$
$$\ldots\#11s_i011\#111s_m11\#11s_m1$$

$$\ldots \#11s_i011\#111s_m11\#$$
$$\ldots \#11s_i011\#111s_m11\#11s_m1\#$$
$$\ldots \#11s_i011\#111s_m11\#1$$
$$\ldots \#11s_i011\#111s_m11\#11s_m1\#1$$
$$\ldots \#11s_i011\#111s_m11\#11s_m1$$
$$\ldots \#11s_i011\#111s_m11\#11s_m1\#1s_m$$
$$\ldots \#11s_i011\#111s_m11\#11s_m1\#$$
$$\ldots \#11s_i011\#111s_m11\#11s_m1\#1s_m\#$$
$$\ldots \#11s_i011\#111s_m11\#11s_m1\#1s_m$$
$$\ldots \#11s_i011\#111s_m11\#11s_m1\#1s_m\#s_m$$
$$\ldots \#11s_i011\#111s_m11\#11s_m1\#1s_m\#$$
$$\ldots \#11s_i011\#111s_m11\#11s_m1\#1s_m\#s_m\#$$
$$\ldots \#11s_i011\#111s_m11\#11s_m1\#1s_m\#s_m\#\#$$
$$\ldots \#11s_i011\#111s_m11\#11s_m1\#1s_m\#s_m\#\#.$$

Formally we give a proof of the theorem. If we have a valid set of sequences describing the acceptance of w by M, using induction on the number of computations we show that there is a partial solution

$$\frac{\#s_0w\#\alpha_1s_1\beta_1\#\alpha_2s_0\beta_2\#\ldots\#\alpha_{n-1}s_{n-1}\beta_{n-1}\#}{\#s_0w\#\alpha_1s_1\beta_1\#\alpha_2s_0\beta_2\#\ldots\#\alpha_{n-1}s_{n-1}\beta_{n-1}\#\alpha_ns_n\beta_n\#}.$$

For $n = 0$, we have

$$\frac{\#}{\#s_0w}.$$

Assuming the statement is true for k, and s_k is not the halt state we have

$$\frac{\#s_0w\#\alpha_1\beta_1\#\alpha_2s_0\beta_2\#\ldots\#\alpha_{n-1}s_{k-1}\beta_{k-1}\#}{\#s_0w\#\alpha_1s_1\beta_1\#\alpha_2s_0\beta_2\#\ldots\#\alpha_{k-1}s_{k-1}\beta_{k-1}\#\alpha_ks_k\beta_k\#}.$$

The next pairs are chosen so the string at the top forms $\#\alpha_ks_k\beta_k\#$ using the rules above. There is at most one pair in the pairs generated by δ that works.

We can thus form

$$\frac{\#s_0w\#\alpha_1s_1\beta_1\#\alpha_2s_0\beta_2\#\ldots\#\alpha_{n-1}s_{k-1}\beta_{k-1}\#\alpha_ks_k\beta_k\#}{\#s_0w\#\alpha_1s_1\beta_1\#\alpha_2s_0\beta_2\#\ldots\#\alpha_{k-1}s_{k-1}\beta_{k-1}\#\alpha_ks_k\beta_k\#\alpha_{k+1}s_{k+1}\beta_{k+1}\#}$$

and we have extended a new partial solution. Since rules generated by δ apply to only one letter, rules $\frac{1}{1}$ and $\frac{0}{0}$ may be needed to produce α_k and β_k. If M

starting with #s_0w reaches a halt state, there is a rule to get from #$\alpha_k s_k \beta_k$# to
#$\alpha_{k+1} s_{k+1} \beta_{k+1}$# otherwise, for some k, there is not a rule and there can be no
match. If, for some k, β_k is a halt state, then as mentioned above, there are rules
to make the upper and lower lists agree.

As already mentioned, if we do not reach the halt state, we cannot have
a match. If we do reach the halt state, we can produce a match. Hence if
the modified Post's Correspondence Problem is decidable, L_0 is decidable.
Therefore the modified Post's Correspondence Problem is undecidable.   □

Example 5.3   Let the Turing Machine

$$M = (\{s_0, s_1, s_2, h\}, \{0, 1\}, \{0, 1, \star, \#\}, \delta, s_0, h)$$

and word 0110 where

$$\delta(s_0, 0) = (s_1, \star, R)$$
$$\delta(s_0, 1) = (s_1, 1, R)$$
$$\delta(s_1, 1) = (s_1, 1, R)$$
$$\delta(s_1, 0) = (s_2, 0, L)$$
$$\delta(s_2, 1) = (s_2, 1, L)$$
$$\delta(s_2, 0) = (s_2, 1, R)$$
$$\delta(s_2, \#) = (h, \#, \#)$$

with corresponding pairs

$$(s_0 0, \star s_1)$$
$$(s_0 1, 1 s_1)$$
$$(s_1 1, 1 s_1)$$
$$\left\{ \begin{array}{l} (0 s_1 0, s_2 00) \\ (1 s_1 0, s_2 10) \\ (\star s_1 0, s_2 \star 0) \end{array} \right.$$
$$\left\{ \begin{array}{l} (0 s_2 1, s_2 01) \\ (1 s_2 1, s_2 11) \\ (\star s_1 1, s_2 \star 1) \end{array} \right.$$
$$(s_2 0, 1 s_2)$$
$$(s_2 \star, h 0).$$

In addition we have pairs

$$(0, 0)$$
$$(1, 1)$$
$$(\#, \#)$$
$$(\star, \star).$$

Our first pair is $(\#, \#s_0 w)$ which produces

$$\#$$
$$\#s_0 010\#.$$

We now use $(s_0 0, \star s_1)$ to get

$$\#s_0$$
$$\#s_0 010\# \star s_1.$$

We then use $(1, 1)$ twice, $(0, 0)$, and $(\#, \#)$ to get

$$\#s_0 0110\#$$
$$\#s_0 0111\# \star s_1 110\#.$$

We next use $(\star, \star)$, $(s_1 1, 1s_1)$, $(1, 1), (0, 0)$, and $(\#, \#)$, to get

$$\#s_0 0110\# \star s_1 110\#$$
$$\#s_0 0110\# \star s_1 110\# \star 1s_1 10\#,$$

again using $(\star, \star)$, $(1, 1)$, $(s_1 1, 1s_1)$, $(0, 0)$, and $(\#, \#)$ we get

$$\#s_0 0110\# \star s_1 110\# \star 1s_1 10\#$$
$$\#s_0 0110\# \star s_1 110\# \star 1s_1 10\# \star 11s_1 0\#.$$

Now using $(\star, \star)$, $(1, 1)$, $(1s_1 0, s_2 10)$, and $(\#, \#)$ we get

$$\#s_0 0110\# \star s_1 110\# \star 1s_1 10\# \star 11s_1 0\#$$
$$\#s_0 0110\# \star s_1 110\# \star 1s_1 10\# \star 11s_1 0\# \star 1s_2 10\#.$$

Using $(\star, \star)$, $(\star s_1 1, s_2 \star 1)$, $(1, 1)$, $(0, 0)$, and $(\#, \#)$ we get

$$\#s_0 0110\# \star s_1 110\# \star 1s_1 10\# \star 11s_1 0\# \star 1s_2 10\# \star s_2 110\#$$
$$\#s_0 0110\# \star s_1 110\# \star 1s_1 10\# \star 11s_1 0\# \star 1s_2 10\# \star s_2 110\#s_2 \star 110\#.$$

Now using $(s_2 \star, h0)$, $(1, 1)$ twice, $(0, 0)$, and $(\#, \#)$, we get

$$\#s_0 0110\# \star s_1 110\# \star 1s_1 10\# \star 11s_1 0\# \star 1s_2 10\# \star s_2 110\#s_2 \star 110\#$$
$$\#s_0 0110\# \star s_1 110\# \star 1s_1 10\# \star 11s_1 0\# \star 1s_2 10\# \star s_2 110\#s_2 \star 110\#h0110\#.$$

Finally, using the pairs containing h, together with $(1, 1)$, $(0, 0)$, and $(\#, \#)$, we get

$$\#s_0 0110\# \star s_1 110\# \star 1s_1 10\# \star 11s_1 0\# \star 1s_2 10\# \star s_2 110\#s_2 \star 110\#h0110\#h$$
$$\#s_0 0110\# \star s_1 110\# \star 1s_1 10\# \star 11s_1 0\# \star 1s_2 10\# \star s_2 110\#s_2 \star 110\#h0110\#h.$$

We can now use the fact that Post's Correspondence Problem is undecidable to solve several other questions about solvability with regard to context-free languages.

Theorem 5.10 *It is undecidable for arbitrary context-free grammars G_1 and G_2 whether $L(G_1) \cap L(G_2) = \emptyset$.*

Proof Let $P \subset \Sigma^* \times \Sigma^*$ be an arbitrary correspondence system with pairs $(u_0, v_0), (u_1, v_1), (u_2, v_2), \ldots, (u_n, v_n)$. In the following, w^{-1} will be w with the letters reversed. For example 1101^{-1} is 1011. Let G_1 be generated by productions

$$
\begin{aligned}
S &\to u_i C v_i^{-1} &&\text{for } i = 1 \text{ to } n. \\
C &\to u_i C v_i^{-1} &&\text{for } i = 1 \text{ to } n. \\
C &\to c.
\end{aligned}
$$

Thus every word in $L(G_1)$ has the form $u_{i_0} u_{i_1} u_{i_2} \ldots u_{i_m} c v_{i_m}^{-1} \ldots v_{i_2}^{-1} v_{i_1}^{-1} v_{i_0}^{-1}$. Let $L(G_2) = \{wcw^{-1} \mid w \in \Sigma^*\}$. Then $w \in L(G_1) \cap L(G_2)$ if and only if $w = u_{i_0} u_{i_1} u_{i_2} \ldots u_{i_m} = v_{i_0} v_{i_1} v_{i_2} \ldots v_{i_m}$ which is a solution to the Post's correspondence system. Hence it is undecidable for arbitrary context-free grammars G_1 and G_2 whether $L(G_1) \cap L(G_2) = \emptyset$. $\square$

Definition 5.8 *A context-free grammar is ambiguous if there are two leftmost generations of the same word.*

Example 5.4 Let $\Gamma = (N, \Sigma, S, P)$ be the grammar defined by $N = \{S, A, B\}$, $\Sigma = \{a, b\}$, and P be the set of productions

$$S \to aSb \quad S \to aA \quad A \to Bb \quad A \to aA \quad B \to Bb \quad B \to \lambda \quad S \to \lambda.$$

Obviously $a^n b^n$ can be generated in two different ways.

Theorem 5.11 *It is undecidable whether an arbitrary context-free grammar is ambiguous.*

Proof Let $P \subset \Sigma^+ \times \Sigma^+$ be an arbitrary correspondence system with pairs $(u_0, v_0)(u_1, v_1), (u_2, v_2), \ldots, (u_n, v_n)$. Let $\alpha_0, \alpha_1, \alpha_2, \ldots, \alpha_n$ be symbols not in Σ^*. We construct two grammars G_1 and G_2 as follows:

$$G_1 = (N_1, \Sigma_a, S_1, P_1)$$

where $N_1 = \{S_1\}$, $\Sigma_a = \Sigma \cup \{\alpha_0, \alpha_1, \alpha_2, \ldots, \alpha_n\}$, and $P_1 = \{S_1 \to \alpha_i S_1 u_i$ for $i = 0, 1, \ldots, n$, and $S_1 \to \lambda\}$.

$$G_2 = (N_2, \Sigma_a, S_2, P_2)$$

where $N_2 = \{S_2\}$, $\Sigma_a = \Sigma \cup \{\alpha_0, \alpha_1, \alpha_2, \ldots, \alpha_n\}$, and $P_2 = \{S_2 \to a_i S_2 v_i$ for $i = 0, 1, \ldots, n$, and $S_2 \to \lambda\}$.

Obviously G_1 and G_2 are not ambiguous.

Let $G = (N, \Sigma_a, S, P)$ where $N = \{S, S_1, S_2\}$ and $P = P_1 \cup P_2 \cup \{S \to S_1, S \to S_2\}$. Obviously if there is a match, one derivation begins with $S \to S_1$ and the other with $S \to S_2$ so G is ambiguous. Conversely if G is ambiguous, then $u_{i_0} u_{i_1} u_{i_2} \dots u_{i_m} = v_{i_0} v_{i_1} v_{i_2} \dots v_{i_m}$ and there is a match. Hence there is a match if and only if the context-free grammar is ambiguous. Therefore it is impossible to determine whether an arbitrary context-free grammar is ambiguous. □

Exercises

(1) Show that the class of Turing acceptable languages is closed under union.
(2) Show that the class of Turing acceptable languages is closed under intersection.
(3) Show that the class of Turing decidable languages is closed under intersection.
(4) Show that the class of Turing decidable languages is closed under union.
(5) Show that the class of Turing decidable languages is closed under concatenation.
(6) Show that the class of Turing decidable languages is closed under Kleene star.
(7) Show that it is an unsolvable problem to determine, for a given Turing machine M, whether there is a string w such that M enters each of the machine's states during its computation of input w.
(8) Show that it is undecidable for any arbitrary context-free grammar Γ whether $\Gamma(M) = \Sigma^*$.
(9) Show that for arbitrary context-free grammars Γ and Γ', it is undecidable whether $\Gamma(L) = \Gamma'(L)$.
(10) Show that there is no algorithm that determinines whether the intersection of languages of two context-free grammars contains infinitely many elements.
(11) Show that there is no algorithm that determines whether the complement of the languages of context-free grammars contains infinitely many elements.

6

A visual approach to formal languages

6.1 Introduction

Formal language theory is overlapped by a close relative among the family of mathematical disciplines. This is the specialty known as **Combinatorics on Words**. We must use a few of the most basic concepts and propositions of this field. A nonnull word, q, is said to be **primitive** if it cannot be expressed in the form x^k with x a word and $k > 1$. Thus, for any alphabet containing the symbols a and b, each of the words a, b, ab, bab, and $abababa$ is primitive. The words aa and $ababab$ are not primitive and neither is any word in $(aba)^+$ other than aba itself. One of the foundational facts of word combinatorics, which is demonstrated here in Section 6.2, is that each nonnull word, w, consisting of symbols from an alphabet Σ, can be expressed in a *unique* way in the form $w = q^n$ where q is a primitive word and n is a positive integer. The uniqueness of the representation, $w = q^n$, allows a useful display of the free semigroup Σ^+, consisting of the nonnull words formed from symbols in Σ, in the form of a Cartesian product, $Q \times N$, where Q *is the set of all primitive words in* Σ^+ and N *is the set of positive integers*. Each word $w = q^n$ is identified with the ordered pair (q, n). This chapter provides the groundwork for investigations of concepts that arise naturally in visualizing languages as subsets of $Q \times N$. In the suggested visualizations, the order structure of N is respected. We regard N as labeling a vertical axis (y-axis) that extends upward only. We regard Q as providing labels for the integer points on a horizontal axis (x-axis) that extends both to the left and right except in the case in which the alphabet is a singleton. When Σ is a singleton the unique symbol in Σ is the only primitive word and $Q \times N$ occupies only the vertical axis.

The set Q of primitive words, over an alphabet having two or more letters, is a rather mysterious language. It is known that Q is not a regular language and that its complement is not a context-free language. (See Exercises 1 and 2 of this section.) At this writing, it is not yet known whether Q itself is context-free. Of course Q is clearly recursive and one can confirm that it is context-sensitive. With Q itself not fully understood, it is surprising that insightful results can be obtained about languages by displaying them in the half plane $Q \times N$. It is as though, in constructing displays above Q, we are building castles on sand. Nevertheless we proceed by taking Q as a *totally structure-less* countable infinite set and we allow ourselves to place Q in one-to-one correspondence with the set of integers (on an x-axis) *in any way we wish* in order to provide the most visually coherent display of the language being treated. One might say that we take advantage of our ability to sprinkle the grains of sand (i.e., primitive words) along the x-axis just as we please.

In the next section adequate tools and exercises are given to allow the introduction of visually coherent displays of languages based on the concept of primitive words.

Exercises

For this set of exercises let $\Sigma = \{a, b\}$ serve as an alphabet and let Q be the set of all primitive words in Σ^+.

(1) Let Q' be the complement of Q in Σ^+ and note that, for every positive integer n, $ab^n aab^n a$ is in Q'. Prove that the language Q' is not regular. Conclude that the language Q also cannot be regular.

(2) Let Q' be the complement of Q and note that, for every positive integer n, $ab^n aab^n aab^n a$ is in Q'. Prove that Q' is not a context-free language. Although complements of context-free languages need not be context-free, there is a subclass of context-free languages, called the **deterministic context-free languages**, for which complements are always context-free. Conclude that Q cannot be a deterministic context-free language.

(3) For each positive integer n, find three primitive words, u, v, and w, for which $uv = w^n$.

(4) Show that both Q and Q' are recursive languages.

(5) Let L be a language that has the property that there is a bound B in N for which for every u in Σ^* there is a word v in Σ^* of length at most B for which uv is in L. Show that L contains an infinite number of primitive words.

(6) Characterize those regular languages that have the property stated in the previous exercise using the intrinsic automaton of L.

6.2   A minimal taste of word combinatorics

Throughout this section, uppercase Σ will denote a nonempty finite set of symbols that will be used as an *alphabet*, while u and v will be reserved to denote *nonnull words* in Σ^+. Uppercase Q will denote *the set of all primitive words* in Σ^+, while p and q will be reserved to denote individual *primitive words*. The following definition for *the division of words* provides a convenient tool for the present exposition: For each pair of words u, v in Σ^+ we define u/v to be the *ordered pair* (n, r) where n is a *nonnegative integer* and $u = v^n r$ with r a suffix of u which does not have v as a prefix. We call r the *remainder* when u is divided by v.

Observe that for each u, v there is only one such pair that meets the conditions in the definition. Note that u is a power of v if and only if $u/v = (n, \lambda)$ in which case $u = v^n$.

Proposition 6.1   *Words u and v are powers of a common word if and only if $uv = vu$. Moreover, when $uv = vu$ both u and v are powers of the word w that occurs as the last nonnull remainder in a sequence of word divisions. The length of this w is the greatest common divisor of the lengths of u and v.*

Proof   Suppose first that $u = w^m$ and $v = w^n$. Then $uv = w^m w^n = w^{m+n} = w^{n+m} = vu$ as required.

Suppose now that $uv = vu$. If $|u| = |v|$ then $u = v$. Otherwise, by the symmetry of the roles of u and v in the hypothesis, we may assume $|v| < |u|$ and we observe that v is a prefix of u. Let $w_0 = u$ and $w_1 = v$. Define successively each word $w_2, w_3, \ldots$, as the remainder, w_i, in the division $w_{i-2}/w_{i-1} = (n_i, w_i)$, stopping when the word $w_k = \lambda$ is obtained. That such a $w_k = \lambda$ must arise is a consequence of the observation that each w_{i-1} is a nonnull prefix of w_{i-2} when w_{i-1} itself is not null. (See Exercise 3 of this section.) We have in succession: $w_{k-2}/w_{k-1} = (n_k, \lambda)$ and w_{k-2} is a power of w_{k-1}, w_{k-3} is a power of w_{k-1}, w_{k-4} is a power of $w_{k-1}, \ldots, w_1(= v)$ is a power of w_{k-1}, $w_0(= u)$ is a power of w_{k-1}. Thus u and v are powers of the common word w_{k-1}. A review of the sequence of divisions confirms that $|w_{k-1}| = \gcd(|u|, |v|)$. $\square$

From the second paragraph of the previous proof we observe that if u and v are powers of a common word, then the length of the *longest* word w for which both u and v are powers of w is the greatest common divisor (gcd) of $|u|$ and $|v|$. Consequently, Proposition 6.1 provides two methods of deciding whether two words u, v are powers of a common word; one can test the equality of uv and vu. Alternatively one can compute $g = \gcd(|u|, |v|)$, test the equality of the prefixes u', v' of length g of u, v, respectively, and if $u' = v'$ then compute u/u' and v/v' to test whether the remainder in each case is λ.

Proposition 6.2 *For each word u in Σ^+ there is a unique pair (q, n), with q in Q and n in N, for which $u = q^n$.*

Proof For each i with $1 \leq i \leq |u|$, let u_i be the prefix of u of length i. Compute successively the u/u_i until a j occurs at which $u/u_j = (m, \lambda)$. Such a j will certainly occur since $u/u_{|u|} = (1, \lambda)$. For the pair (u_j, m) we have u_j in Q. Consequently $u = u_j^m$ has the required form. Suppose now that $u = p^m = q^n$, where both p and q are in Q.

$$uu = p^m p^m = pp^{m-1} p^m = pp^m p^{m-1} = pup^{m-1} = pq^n p^{m-1} = (pq)q^{n-1} p^{m-1},$$

and

$$uu = q^n q^n = qq^{n-1} q^n = qq^n q^{n-1} = quq^{n-1} = qp^m q^{n-1} = (qp)p^{m-1} q^{n-1}.$$

Thus $pq = qp$ and by Proposition 6.1, p and q must be powers of a common word. Since each is primitive, $p = q$ and then also $m = n$, as required for uniqueness. $\square$

For each word u the unique primitive word q for which u is a power of q will be called *the primitive root of u* and will be denoted $rt(u)$. Thus Proposition 6.1 may be rephrased: $uv = vu$ *if and only if $rt(u) = rt(v)$*. Note that for each word u, $u = rt(u)^n$ for a unique n in N. Thus for each positive integer m, $u^m = rt(u)^{nm}$. Thus $rt(u^m) = rt(u)$ for each word u and each positive integer m. The exponent n in the unique representation $u = rt(u)^n$ will be called *the exponent of u*.

Propositions 6.1 and 6.2 are the bedrock of the theory of word combinatorics and should become familiar tools for anyone studying formal languages. They contain only the information required to begin the discussion of language visualization. The additional information about word combinatorics that is required in the algorithmics of visualization is given in Section 8 of this chapter. For further study of the fascinating but subtle mathematics of word combinatorics see [36] [35] [28] and [13].

Exercises

(1) Let $\Sigma = \{a, b, c\}$ serve as alphabet. Let

$$u = abcabcabcabcabcabcabcabcabcabc$$

and

$$v = abcabcabcabcabcabc.$$

Given that u and v commute, carry out the steps of the procedure given in the second paragraph of the proof of Proposition 6.1 for finding the longest word w for which u and v are powers of w. Give the finite sequence of the words $w_0, w_1, w_2, \ldots, w_k$ that arises in this computation.

(2) Determine whether the words u and v as in the previous exercise are powers of a common string w by each of the two methods stated in the paragraph following Proposition 6.1. Note that the second method uses Euclid's number theoretic algorithm for finding greatest common divisors of integers and produces the *longest* such w when u and v are powers of a common string. The first method given produces the *shortest* such w when u and v are powers of a common string.

(3) The proof of Proposition 6.1 contains the following assertion: "That such a $w_k = \lambda$ must arise is a consequence of the observation that each w_{i-1} is a nonnull prefix of w_{i-2} when w_{i-1} itself is not null." Prove the observation that each w_{i-1} is a nonnull prefix of w_{i-2} when w_{i-1} itself is not null.

(4) Observe that the length function $|\ | : \Sigma^+ \to N$ is a semigroup homomorphism that maps Σ^+ onto the additive semigroup N. Let u and v be any two nonnull words in Σ^+ for which $uv = vu$. Show that the restriction of the length function to the subsemigroup $\{u, v\}^+$ is a semigroup homomorphism $|\ | : \{u, v\}^+ \to N$ that maps $\{u, v\}^+$ *one-to-one* into the additive semigroup N. What fails in your argument if $uv \neq vu$?

6.3 The spectrum of a word with respect to a language

With each language L and each nonnull word w we define a subset of the positive integers N called **the spectrum**, $\mathrm{Sp}(w, L)$, of w **with respect to** L : $\mathrm{Sp}(w, L) = \{n \in N : w^n \text{ is in } L\}$. In the display of L in the half plane $Q \times N$ the column at each primitive word, q, displays the spectrum of q. In fact, the display of the spectra of the words in Q constitutes the representation of L within $Q \times N$. It is the spectra of the primitive words that are of primary concern (since the spectrum of w^n can be read directly from the spectrum of w). However, in Section 6.9 the value of defining the spectra of the nonprimitive words along with the primitive words is justified.

It is convenient to classify spectra into five qualitatively distinct categories. The spectrum of a word with respect to a language L may be the *empty* set, a *finite* set, a *cofinite* set, the entire set N, or an *intermittent* set. When the spectrum is the entire set N we say that the spectrum is *full*. Recall that a set is cofinite if it has a finite complement. When a spectrum is empty it is also finite and when a spectrum is full it is also cofinite. By an intermittent spectrum we mean a spectrum that is neither finite nor cofinite. Note that if $\mathrm{Sp}(w, L)$ is intermittent then, for every positive integer n, there are integers $i > n$ and $j > n$ for which $w^i \in L$ and w^j is not in L.

Each of the five cases is easily illustrated using the one letter alphabet $\Sigma = \{a\}$: $\mathrm{Sp}(a,$ the empty language $\emptyset)$ is empty. $\mathrm{Sp}(a, \{a, aaa\})$ is the finite set $\{1, 3\}$. $\mathrm{Sp}(a, a \vee aaa^+)$ is the cofinite set $\{n \in N : n = 1 \text{ or } n \geq 3\}$. $\mathrm{Sp}(a, a^+)$ is the full set N. $\mathrm{Sp}(a, (aa)^+)$ is intermittent, being $\{n \in N : n \text{ is even}\}$. The single letter a is the only primitive word for the alphabet $\Sigma = \{a\}$. Spectra of nonprimitive words for these same languages are illustrated: $\mathrm{Sp}(aa, \emptyset) = \emptyset$, $\mathrm{Sp}(aaa, \{a, aaa\}) = \{1\}$, $\mathrm{Sp}(aa, a \vee aaa^+) = \{n \in N : n \geq 2\}$, $\mathrm{Sp}(aaa, a^+)$ is full, and $\mathrm{Sp}(aaa, (aa)^+) = \{n \in N : n \text{ is even}\}$. Note that, in the case of one letter as alphabets, such as $\Sigma = \{a\}$, the distinction between a language and the spectrum of the letter a is somewhat artificial. Consider now the two letter alphabet $\Sigma = \{a, b\}$. The spectrum of each word w with respect to the context-free language $L = \{w \in \Sigma^+ : a \text{ and } b \text{ occur equally often in } w\}$ is either empty or full; $\mathrm{Sp}(w, L)$ is full if w is in L and empty otherwise. For the regular language $L = (aa)^+ \vee (bbb)^+$, $\mathrm{Sp}(a^n, L)$ is full if n is even and intermittent if n is odd. $\mathrm{Sp}(b^n, L)$ is full if n is divisible by three and intermittent otherwise. Finally, $\mathrm{Sp}(w, L) = \emptyset$ if both a and b occur in w.

Exercises

(1) Let L be a *regular* language in Σ^+ and let N be represented in tally notation $N = |^+$. Show that, for any word w in Σ^+, $\mathrm{Sp}(w, L)$ is a regular language in $N = |^+$.

(2) Let L be a *context-free* language in Σ^+ and let N be represented in tally notation. Show that, for any word w in Σ^+, $\mathrm{Sp}(w, L)$ is a regular language in $N = |^+$.

(3) Let Σ be a finite alphabet that contains at least two symbols. Let w be one specific *fixed* word in Σ^+. For each of the following sets state whether the set is countably infinite or uncountably infinite:

    (a) $\{L : L \text{ is a language contained in } \Sigma^+\}$,

    (b) $\{\mathrm{Sp}(w, L) : L \text{ ranges through } all \text{ the languages in } \Sigma^+\}$,

    (c) $\{\mathrm{Sp}(w, L) : L \text{ ranges through all the } regular \text{ languages in } \Sigma^+\}$, and

    (d) $\{\mathrm{Sp}(w, L) : L \text{ ranges through all the } context\text{-}free \text{ languages in } \Sigma^+\}$.

6.4   The spectral partition of Σ^+ and the support of L

Let L be a language contained in Σ^+. This language L provides an equivalence relation, $\sim$, defined for words u and v in Σ^+, by setting $u \sim v$ provided u and v have identical spectra, i.e., $\mathrm{Sp}(u, L) = \mathrm{Sp}(v, L)$. We call the partition provided by $\sim$ the **spectral partition**, $P(L)$, of Σ^+ **induced by** L. This partition is a fundamental tool for the present study. In Section 6.7 it is observed that, when L is regular, $P(L)$ consists of a finite number of constructible regular languages. Using a refinement of $P(L)$ and Theorem 6.1 gives a precise view of L, within $Q \times N$, when L is a *regular language*. The spectral partitions determined by the languages discussed in Section 6.3 are given next as examples.

Let $\Sigma = \{a\}$. For the language $L = \Sigma^+$, the spectrum of every word in Σ^+ is full. Consequently $P(L) = P(\Sigma^+)$ consists of a single class, i.e., $P(\Sigma^+) = \{\Sigma^+\}$. For the empty language, $\emptyset$, the spectrum of every word in Σ^+ is $\emptyset$. Thus $P(\emptyset)$ also consists of the single class $\{\Sigma^+\}$. For $L = \{a, aaa\}$, $P(L) = \{\{a\}, \{aaa\}, \Sigma^+\backslash L\}$. For $L = a \vee aaa^+$, $P(L) = \{\{a\}, \{aa\}, aaa^+\}$. For $L = (aa)^+$, $P(L) = \{\{a^n : n \text{ is odd}\}, \{a^n : n \text{ is even}\}\}$. Now let $\Sigma = \{a, b\}$. For $L = \{w \text{ in } \Sigma^+ : a \text{ and } b \text{ occur equally often in } w\}$, $P(L) = \{L, \Sigma^+\backslash L\}$. For $L = (aa)^+ \vee (bbb)^+$, $P(L) = \{L, a(aa)^*, b(bbb)^* \vee bb(bbb)^*, \Sigma^*ab\Sigma^* \vee \Sigma^*ba\Sigma^*\}$.

For visualizing a language, L, within $Q \times N$, the spectra of the primitive words in Σ^+ provide the whole picture. If desired, the spectrum of a nonprimitive word, q^n, can be obtained from the spectrum of its primitive root, q. In fact, for the task at hand here, there is little motive for interest in the spectra of individual nonprimitive words. For each equivalence class, C, in $P(L)$ we are actually only interested in $C \cap Q$. The single reason for providing the definition of the spectra of nonprimitive words is that each resulting spectral class, C, can often provide satisfactory access to the crucial set of primitive words $C \cap Q$. The first three crucial questions we ask about a set $C \cap Q$ are: (a) Is $C \cap Q$ empty? (b) If not, is $C \cap Q$ infinite? (c) If $C \cap Q$ is finite, can its elements be listed? These questions are answered for the languages discussed in the previous paragraph in order to provide examples.

For a one letter alphabet, $\Sigma = \{a\}$, the letter itself is the only primitive word. Consequently for any language L contained in Σ^+, $C \cap Q$ is empty for each C other than the one containing the letter a. Now let $\Sigma = \{a, b\}$. For $L = \{w \text{ in } \Sigma^+ : a \text{ and } b \text{ occur equally often in } w\}$, each of the two classes in $P(L) = \{L, \Sigma^+\backslash L\}$ contains an infinite number of primitive words. For $L = (aa)^+ \vee (bbb)^+$, we previously obtained $P(L) = \{L, a(aa)^*, b(bbb)^* \vee bb(bbb)^*, \Sigma^*ab\Sigma^* \vee \Sigma^*ba\Sigma^*\}$. For these four spectral classes we have:

$L \cap Q$ is empty; $(a(aa)^*) \cap Q = \{a\}$; $(b(bbb)^* \vee bb(bbb)^*) \cap Q = \{b\}$; and $(\Sigma^* ab\Sigma^* \vee \Sigma^* ba\Sigma^*) \cap Q$ is infinite.

For each language L contained in Σ^+, the set $\mathrm{Su}(L) = \{q \in Q : \mathrm{Sp}(q, L)$ is not empty$\}$ will be called the **support** of L. For a one letter alphabet, $\Sigma = \{a\}$, the support of each nonempty language L is Σ itself. Now let $\Sigma = \{a, b\}$. For the language $L = \{w$ in $\Sigma^+ : a$ and b occur equally often in $w\}$, $\mathrm{Su}(L)$ is the infinite set $L \cap Q$. For $L = (aa)^+ \vee (bbb)^+$, $\mathrm{Su}(L)$ is the finite set $\{a, b\}$. The cardinality of the support of a language is of special significance for the investigations introduced here. When a support is finite, the specific primitive words in the support are desired.

Exercises

(1) For $\Sigma = \{a, b\}$ and $L = \{(ab^m)^n : m, n \in N\}$:
    (a) determine the spectrum of each of the words ab, $abbabbabb$, $ababb$; state whether each spectrum is empty, finite, cofinite, full, or intermittent;
    (b) determine the spectral partition $P(L)$; and
    (c) determine the support $\mathrm{Su}(L)$ and state whether it is a regular language.

(2) Let $\Sigma = \{a, b\}$ and $L = \{a^n b^n : n \in N\}$.
    (a) Confirm that the spectrum of each word in Σ^+ is either $\emptyset$ or $\{1\}$.
    (b) Determine $P(L)$ and $\mathrm{Su}(L)$.
    (c) For the language LL, determine the spectra of ab, $abab$, and $ababab$.
    (d) Describe $P(LL)$ and $\mathrm{Su}(LL)$.

6.5 Visualizing languages

In order to spell out the visualization of a language L within $Q \times N$, we begin with the usual x–y plane with each point having associated real number coordinates (x, y). *We use only the upper half plane,* $\{(x, y) : y > 0\}$. With each integer i and each positive integer n we associate the unit rectangle

$$R(i, n) = \{(x, y) : i - 1 < x \le i, n - 1 < y \le n\}.$$

In this way the upper half plane is partitioned into nonoverlapping unit squares $\{R(i, n) : i$ an integer, $n \in N\}$. To visualize a specific language L in Σ^+ we first identify the set Q with the set Z of integers using any chosen bijection $B : Q \to Z$. (The bijection B is chosen only after a study of the spectral partition of the specific language L has been made, as illustrated below.) Once the bijection B is chosen, each word q^n in Σ^+ is associated with (figuratively, "placed on") the unit square $R(B(q), n)$. Finally, the language L is visualized by defining, using B, a **sketch function** $S : \{R(B(q), n) : q \in Q, n \in N\} \rightarrow$ {Black, White} for which $S(R(B(q), n)) = $ Black if q^n is in L and White otherwise. For each given language L and each bijection B, the resulting sketch function is said to provide a **sketch** of the language L. By the sketch we mean the *image* of the sketch function that provides it. Thus we regard the sketch as a half plane in which each of the unit squares is either black or white. Since there are many possible choices for B, there may be many possible sketches of L. For many languages, coherent sketches can be given by basing the choice of the bijection B on a determination of the spectral decomposition of the language. Examples follow for which we use the alphabet $\Sigma = \{a, b\}$. These examples suggest several new formal language concepts that we believe are worthy of theoretical development. Each definition given in this section follows immediately below one or more examples that illustrate or clarify the concept being defined.

Example 6.1　For $L = \{w$ in $\Sigma^+ : a$ and b occur equally often in $w\}$, each of the two spectral classes in $P(L) = \{L, \Sigma^+\backslash L\}$ contains an infinite number of primitive words. The spectrum of each word in L is full and the spectrum of each word in $\Sigma^+\backslash L$ is empty. Let B be any bijection for which $B(L \cap Q) = \{i \in Z : i \leq 0\}$ and $B((\Sigma^+\backslash L) \cap Q) = \{i \in Z : i \geq 1\}$. The sketch provided by this choice of B gives a black left quadrant and a white right quadrant. The support of this language is the infinite set $L \cap Q$.

Definition 6.1　*A language L is **cylindrical** if, for each word w in Σ^+, $\mathrm{Sp}(w, L)$ is either empty or full.*

The language L of Example 6.1 is cylindrical. There are numerous "naturally occurring" examples of cylindrical languages: The fixed language $L = \{w \in \Sigma^+ : h(w) = w\}$ of each endomorphism h of Σ^+ is a cylindrical regular language and so is the stationary language of each such endomorphism [16] [15]. Retracts and semiretracts [16][10][3][1] of free monoids are cylindrical languages. Investigations of various forms of periodicity in the theory of Lindenmayer systems have led to additional examples of cylindrical languages [24][26].

Example 6.2　For $L = \{aa, aaa, aaaa, aaaaaa, bbb, bbbb, ababab\}$ only three primitive words have nonempty spectra: a, b, and ab. Let B be any bijection for which $B(a) = 1$, $B(b) = 2$, and $B(ab) = 3$. The sketch provided by such a B gives a half plane that is white except for the three columns above the three primitive words a, b, and ab. The column above a reads, from the bottom up, white, black, black, black, white, black, and white thereafter. The column above b reads white, white, black, black, and white thereafter. The column above ab reads white, white, black, and white thereafter. The support of this language is the finite set $Su(L) = \{a, b, ab\}$.

Example 6.3 For $L = \{(a^m b)^n : m, n \in N, m \geq n\}$, the support of L is $Su(L) = \{a^m b : m \in N\}$. Let B be any bijection for which, for each m in N, $B(a^m b) = m$. The sketch provided by such a B gives a white left quadrant. The right quadrant is white above a sequence of black squares ascending upward at 45 degrees and black below this sequence of squares. The support of this nonregular language is the infinite regular language $a^+ b$.

Definition 6.2 *A language L is **bounded above** if, for each word w in Σ^+, $Sp(w, L)$ is finite. A language L is **uniformly bounded above** if it is bounded above and there is a positive integer b for which, for each w in Σ^+ and each n in $Sp(w, L)$, $n \leq b$.*

Any finite language, such as the one given in Example 6.2, is necessarily uniformly bounded above. An infinite language may also be uniformly bounded above (Exercise 4, below in this section) or bounded above without a uniform bound, as illustrated in Example 6.3.

Example 6.4 For $L = \{a, aaa, aaaaa\} \cup b(a \vee b)^*$, each word that begins with a b has a full spectrum and each word that begins with an a and contains a b has an empty spectrum. Let B be any bijection for which $B(a) = 1$; $B(Q \cap b(a \vee b)^*) = \{n \in N : n \geq 2\}$. The sketch provided by such a B gives a white left quadrant and a right quadrant that is black except for the column above a which reads black, white, black, white, black, and white thereafter. The support of this regular language is the infinite nonregular set $Su(L) = L \cap Q$.

Example 6.5 For $L = \{(a^m b)^n : m, n \in N, m \text{ odd}, m \geq n\} \cup \{(a^m b)^n : m, n \in N, m \text{ even}, m \leq n\}$, the support of L is $Su(L) = a^+ b$. Let B be any bijection for which, for each m in N, $B(a^m b) = m$. The sketch provided by such a B gives a white left quadrant. The right quadrant has a sequence of black squares ascending upward at 45 degrees. For each odd positive integer m, $(a^m b)^n$ is black for $n \leq m$ and white for $n > m$. Whereas, for each even positive integer m, $(a^m b)^n$ is white for $n < m$ and black for $n \geq m$.

Definition 6.3 *A language L is **eventual** if, for each word w in Σ^+, $Sp(w, L)$ is either finite or cofinite. The language L is **uniformly eventual** if there is an m in N for which, for each word w in Σ^+, either $Sp(w, L) \subseteq \{n \in N : n < m\}$ or $Sp(w, L) \supseteq \{n \in N : n \geq m\}$.*

The language of Example 6.4 is uniformly eventual. The language of Example 6.5 is eventual but not uniformly eventual. Note that each cylindrical language is uniformly eventual (where any n in N may be taken as the uniform bound). Note also that each language that is (uniformly) bounded above is (uniformly) eventual. *Every uniformly eventual language is the symmetric difference of a cylindrical language and a language that is uniformly bounded above* (Exercise 2, below in this section). Each noncounting language [31] is uniformly eventual as was pointed out in [15] where the concept of an eventual language was first introduced.

Example 6.6   For $L = aa \vee aaa \vee (aabaab)^+ \vee (ababab)^+ \vee b(a \vee b)^*$, each word that begins with a b has a full spectrum. Each primitive word that begins with an a has an empty spectrum except for the primitive words a, aab, and ab. Let B be any bijection for which $B(a) = 1$, $B(aab) = 2$, $B(ab) = 3$, and $B(b(a \vee b)^* \cap Q) = \{n \in N : n \geq 4\}$. The sketch provided by such a B gives a white left quadrant and a right quadrant that is black except for three columns. The column above a reads: white, black, black, and white thereafter. The columns above aab and ab are both intermittent with the first having period two and the second having period three. Therefore $Su(L) = \{a, aab, ab\} \cup (Q \cap (b\Sigma^*))$.

Definition 6.4   *A language L is **almost cylindrical** (respectively, **almost bounded above, almost uniformly bounded above, almost eventual, almost uniformly eventual**) if it is the union of a language with finite support and a language that is cylindrical (respectively, bounded above, uniformly bounded above, eventual, uniformly eventual).*

The language of Example 6.6 is almost cylindrical and therefore also almost uniformly eventual. The uniformly eventual language of Example 6.4 is almost cylindrical. The language of Exercise 3 in this section below, is uniformly eventual, almost uniformly bounded above, and also almost cylindrical. The union of the languages of Exercises 3 and 4 in this section below is uniformly eventual and almost uniformly bounded above but not almost cylindrical. The union of the languages of Examples 6.5 and 6.6 is almost eventual, but not almost uniformly eventual.

John Harrison provided the first application of the concept of an almost cylindrical language in [14].

If humor can be tolerated, we may say that the freedom we allow in choosing the bijections B for determining our language sketches can be supported with the slogans: "All Primitives Were Created Equal", "End Domination by Alphabetical Symbols", and "Power to the Primitives!"

Exercises

(1) Regular languages that have a given property often have the uniform version of the property:

    (a) Show that every regular language that is bounded above is uniformly bounded above.

    (b) Show that every regular language that is almost eventual is uniformly almost eventual. (Both parts of this exercise may be easier after reading Section 6.10.)

(2) Show that each uniformly eventual language L is the symmetric difference of a cylindrical language and a language that is uniformly bounded above.

(3) Let $L = a \vee aaa \vee (ab)^+ \vee b^+$. Describe the spectrum of each word in Q. Find $P(L)$ and $Su(L)$. Choose a bijection $B : Q \to Z$ which will provide a coherent sketch of L. Describe this sketch.

(4) Let $L = \{a^n b^n : n \in N\}$. Describe the spectrum of each word in Q. Find $P(L)$ and $Su(L)$. Choose a bijection $B : Q \to Z$ which will provide a coherent sketch of L. Describe this sketch.

(5) Let $L = (\Sigma \Sigma)^+$. Describe the spectrum of each word in Σ^+. State whether each spectrum is empty, finite, cofinite, full, or intermittent. Find $P(L)$ and $Su(L)$. Choose a bijection $B : Q \to Z$ which will provide a coherent sketch of L. Describe this sketch.

(6) Let $L = (ab^+ ab^+)^+$. Find $P(L)$ and $Su(L)$. Choose a bijection $B : Q \to Z$ which will provide a coherent sketch of L. Describe this sketch.

(7) Let $L = ((ab^+)^6)^+$. Find $P(L)$ and $Su(L)$. Choose a bijection $B : Q \to Z$ which will provide a coherent sketch of L. Describe this sketch.

6.6   The sketch parameters of a language

Each sketch of a language L in Σ^+ is given by a sketch function S that is determined entirely by L and the choice of a bijection $B : Q \to Z$. Given two sketches of the same language L, each can be obtained from the other by an appropriate permutation of columns appearing in the sketches. Mathematically, distinguishing between different sketches of the same language L is rather artificial. The distinctions have been made because we prefer the more visually coherent sketches to those that are less visually coherent. The class of all sketches of a given language is determined by any one of its members. Observe that the sketches of a language L are determined by what we call the **sketch parameters of** L that we define as follows: There is one sketch parameter for each spectral class C that contains at least one primitive word. The parameter associated with such a C is the ordered pair consisting of the spectrum of any primitive word q in C and the cardinal number of $C \cap Q$. This sketch parameter is therefore the ordered pair $(\text{Sp}(q), K)$ where q is in $C \cap Q$ and K is the cardinal number of $C \cap Q$. In the discussion of the examples that follows, the cardinal number of N, i.e. the denumerable infinite cardinal, is denoted by the symbol ∞.

For Example 6.1 of Section 6.5, there are only two sketch parameters, (N, ∞) and $(\emptyset, \infty)$. For Example 6.2 of the same section, there are four sketch parameters, $(\{2, 3, 4, 6\}, 1)$, $(\{3, 4\}, 1)$, $(\{3\}, 1)$, and $(\emptyset, \infty)$. Example 6.3 has parameters $(\emptyset, \infty)$ and, for each n in N, $(\{m : m \leq n\}, 1)$. Example 6.4 has parameters $(\{1, 3, 5\}, 1)$, (N, ∞), and $(\emptyset, \infty)$. Example 6.5 has parameters as follows: for each m in N with m odd, $(\{n \in N : m \geq n\}, 1)$; for each m in N with m even, $(\{n \in N : m \leq n\}, 1)$; and finally $(\emptyset, \infty)$. Example 6.6 has five parameters $(\{2, 3\}, 1)$, $(\{2n : n \in N\}, 1)$, $(\{3n : n \in N\}, 1)$, (N, ∞), and $(\emptyset, \infty)$.

We say that two languages are **sketch equivalent** if they can be represented by a common sketch. For example, the context-free language L of Example 6.1 is sketch equivalent to the regular language $b(a \vee b)^*$ since each can be represented by a sketch that has a black left quadrant and a white right quadrant. Similarly the context-free language of Exercise 4 of Section 6.5 is sketch equivalent to the regular language ba^* since each can be represented by a sketch that has a white left quadrant and a right quadrant that is white except for one horizontal black stripe at $n = 1$. Since the sketch parameters of a language determine the class of all possible sketches of a language, *two languages are sketch equivalent if and only if they have the same sketch parameters*. Consequently if L and L' are languages for which the sketch parameters can be determined, then one may be able to decide whether L and L' are sketch equivalent by comparing the sketch parameters of L and L'. This will certainly be the case if one of the languages has only finitely many sketch parameters. In Section 6.10, it is shown that every regular language has only finitely many sketch parameters and that they can be calculated.

Open Ended Exercise Investigate the sketch parameters of $QQ = \{pq : p, q \in Q\}$.

Open Ended Exercise Which *sets* of sketch parameters can occur as the set of sketch parameters of a language L? This question becomes more interesting when L is required to be regular. The regular case might be considered again after reading one or more of the remaining sections.

6.7 Flag languages

Each language L is recognized by its intrinsic automaton $M(L)$. The concept of the recognition of a language by an automaton is thoroughly classical, at least for the regular languages. A concise presentation for arbitrary languages has been included in Chapter 3. In this chapter we apply $M(L)$ only to the study of the spectra of regular languages, although applications may be possible in additional contexts. The *notation* of Chapter 3 is used to give a perfectly explicit discussion of the sketches of regular languages.

Assume now that L is a regular language in Σ^+ and that its recognizing automaton $M(L)$ has m states. With each word w in Σ^+ we associate a finite sequence of states of $M(L)$ in the following way: Consider the infinite sequence of states, $\{[w^n] : n$ a nonnegative integer$\}$. Since $M(L)$ has only m states, there is a least nonnegative integer i for which there is a positive integer j for which $[w^i] = [w^{i+j}]$. Let r be the least positive integer for which $[w^i] = [w^{i+r}]$. We call the sequence $\{[w^n] : 0 \le n < i + r\}$ *the* **flag $F(w)$ of the word** w. The **length** *of* $F(w)$ is $i + r$. Since $M(L)$ has only m states, the maximum length of the flag of any word is m. The collection of distinct flags $\{F(w) : w \in \Sigma^+\}$ associated with a regular language L is necessarily finite. By a **flag F of the language** L we mean a sequence of states that constitutes the flag, relative to L, of some word in w in Σ^+. With each flag F of L we associate the language $I(F) = \{w \in \Sigma^+ : F(w) = F\}$. We call $I(F)$ **the language of the flag F.** For each flag $F = \{s_j : 0 \le j \le k\}$, where the s_i denote the states in F, we have

$$I(F) = \bigcap_j \{L(s_j, s_{j+1}) : 0 \le j \le k - 1\}$$

where each $L(s_j, s_{j+1})$ is the language that consists of all words x in Σ^+ for which $s_j x = s_{j+1}$. Since each of the languages $L(s_j, s_{j+1})$ is regular, *each flag language is regular*. The great value of the flag languages, for regular L, is that they constitute a *finite partition* of Σ^+ into equivalence classes each of which is *a nonempty regular language*. The flag partition of Σ^+ into the flag languages determined by L is denoted $P'(L)$.

Open Ended Exercise In the theory of Abelian Groups the concept of torsion plays a fundamental role [8]. Can this suggest a worthwhile concept of torsion for language theory? A *first attempt* might begin with the tentative definition: A word w in Σ^+ is a **torsion word with respect to a language** L if the flag of w in $M(L)$ is finite. If this definition is used then, for each *regular* L, all words in A^+ would be torsion words with respect to L. The torsion words with respect to the context-free language $L = \{w$ in $\Sigma^+ : a$ and b occur equally often in $w\}$ would be the words in $\{\mu v : \mu \in \Sigma^*, v \in L\}$.

6.8   Additional tools from word combinatorics

This section contains three additional propositions on word combinatorics that are needed for the algorithmics of the next section. Two words x and y are said to be **conjugates** of one another if they possess factorizations of the form $x = uv$ and $y = vu$. From the next proposition it follows that *conjugates have the same exponent*, which includes the information that *the conjugates of primitive words are primitive*. This last fact, that conjugates of primitives are primitive, is applied *many* times in Section 6.9.

Proposition 6.3   *If $uv = p^n$ then $vu = q^n$ with q a conjugate of p.*

Proof   Since $uv = p^n$, we may assume that $p = u''v'$ where $u = p^i u''$ and $v = v'p^j$ with i and j nonnegative integers for which $i + j = n - 1$. For $q = v'u''$ we have

$$q^n = (v'u'')^n = v'(u''v')^{n-1}u'' = v'p^{n-1}u'' = v'p^j p^i u'' = vu.$$

$\square$

Lemma 6.1   *Let v be a word for which $vv = xvy$ with x and y nonnull, then $v = xy = yx$.*

Proof   Since $|v| = |x| + |y|$ and v has x as a prefix and y as a suffix, $v = xy$. Then $vv = xvy$ gives $xyxy = xxyy$ and by cancellation $yx = xy$. $\square$

Proposition 6.4   *If u^i and v^j, with $i, j \geq 2$, have a common prefix of length $|u| + |v|$ then u and v are powers of a common word.*

Proof   By the symmetry of u and v in the hypothesis, we may assume $|u| \geq |v|$. Then v is a prefix of u and $u = v^n x$ where $u/v = (n, x)$. The v that occurs as the prefix of the *second* u, in the series of us concatenated to form u^i, occurs also as a factor of the product of the two vs that occur as the $(n + 1)$st and the $(n + 2)$nd vs concatenated to form v^j. This provides a factorization of the form $vv = xvy$. By Lemma 6.1 and Proposition 6.1, x and y are powers of a common word and therefore so are $v = xy$ and $u = (xy)^n x$. $\square$

Proposition 6.5   *Let u and v be words that are not powers of a common word. For each n in N either $u^n v$ is primitive or $u^{n+1} v$ is primitive. Consequently the set $Q \cap u^+ v$ is infinite.*

Proof   If both $u^n v$ and $u^{n+1} v$ fail to be primitive then, by Proposition 6.3, the conjugate $u^n vu$ of $u^{n+1} v$ also fails to be primitive and we have $u^n v = p^i$ and $u^n vu = q^j$ with p, q primitive and $i, j \geq 2$. Then $p^{2i} = u^n vu^n v$ and $q^j = u^n vu$ have the common initial segment $u^n vu$ which has length

$|u^n vu| = (1/2)|u^n vu| + (1/2)|u^n vu| > (1/2)|u^n v| + (1/2)|u^n vu| > |p| + |q|.$
By Proposition 6.4, p and q are powers of a common word and, since they are primitive, $p = q$. We then have $u^n v = p^i$ and $u^n vu = p^j$, which gives the contradiction: $u = p^{j-i}$ and $v = p^{i-n(j-i)}$. Finally, since at least one member of each pair from the infinite collection of pairs $\{u^k v, u^{k+1} v\}$ must be primitive, the set $Q \cap u^+ v$ is infinite. $\qquad\square$

Exercises

(1) Provide an alternative proof of Proposition 6.2 using Proposition 6.4.
(2) Provide an alternative proof of Proposition 6.2 using Lemma 6.1.
(3) Let Σ be an alphabet containing the symbols a and b. Let u be any word in Σ^+. Show that at least one of ua and ub must be primitive.
(4) Let u and v be in A^+. Suppose that, for some n in N, no word in the set $\{u^k v \mid k \geq n\}$ is primitive. Prove that $uv = vu$. Can you prove this using only Lemma 6.2 without using either Proposition 6.4 or Proposition 6.5?

6.9 Counting primitive words in a regular language

In order to construct the sketch parameters of a language L we will need to determine the cardinal number of the set $C \cap Q$ for each spectral class C of L. The conceptually simple instructions for finding the cardinal of each set $L \cap Q$ for *any regular language* L are given next and followed by a justification that is a simplified version of a proposition provided by M. Ito, M. Katsura, H. J. Shyr and S. S. Yu in [25].

The Counting Procedure Let A be an alphabet with at least two symbols and let L be a regular language contained in A^+. Let $n \geq 2$ be the number of states of the automaton $M(L)$ that recognizes L and let $B = 4n$. Begin testing the primitivity of words in L of length $\leq B$. As the testing progresses maintain a list of all primitive words found thus far. If a primitive word p with $|p| \geq n$ is encountered, STOP with the information that $|L \cap Q| = \infty$. Otherwise continue the testing and the listing process for words in L of length $\leq B$ until either a primitive word p with $|p| \geq n$ is encountered, or all the words in L of length $\leq B$ have been tested. If this procedure has not STOPPED with the information that $|L \cap Q| = \infty$, then the final list of primitive words found is the complete list of all primitive words in L. Such a list will be finite and may be empty.

This counting procedure is justified by the following result:

Theorem 6.1 *Let Σ be an alphabet with at least two symbols and let L be a regular language contained in Σ^+. Let $n \geq 2$ be the number of states of the automaton $M(L)$ that recognizes L and let $B = 4n$. Then: (1) $L \cap Q$ is empty if it contains no primitive word of length $\leq B$; (2) $L \cap Q$ is infinite if it contains a primitive word of length $\geq n$; and (3) if $L \cap Q$ is infinite then it contains a primitive word p with $|p| \leq B$.*

Proof Note first that (1) will follow immediately once (2) and (3) are proved: Suppose $L \cap Q$ contains no primitive word of length $\leq B$. Then, if $L \cap Q$ contained any primitive word at all, that word would have length $> B$. Then $L \cap Q$ would be infinite by (2) and would contain, by (3), a primitive word p for which $|p| \leq B$ contradicting the original supposition. Next we prove (2).

We consider two distinct cases: Suppose first that for every state $[u]$ in $M(L)$, there is a word v in Σ^+ for which $[uv]$ is a final state. Since $M(L)$ has only n states it follows that there is such a word v with $|v| \leq n - 1$. Let a and b be two distinct symbols in Σ. For every integer $i \geq n - 1$ there is a word v of length $\leq n - 1$ for which $a^i b v$ is in L. Each such $a^i b v$ is primitive and is therefore in $Q \cap L$. Consequently $Q \cap L$ is infinite as was to be proved. Surprisingly, perhaps, for this case we have a stronger version of (3) since a word $w = a^{n-1} b v$ lies in $Q \cap L$ and $n \leq |w| \leq 2n - 1 < B$. (Exercises 5 and 6 of Section 6.1 contain related concepts.)

Now suppose that $M(L)$ has a state g for which $[gv]$ is not final for any word v in Σ^+. Such a state g is often called a *dead* state. Suppose that w is in $L \cap Q$ and $|w| = r \geq n$. As w is read by $M(L)$, a walk is made from the initial state to a final state and this walk enters r states after leaving the initial state. This walk does not enter g. Since this walk involves a sequence of $r + 1 \geq n + 1$ states there must be a repetition of states among the last n states in the list. This gives a factorization $w = uxv$ for which $[u] = [ux]$ where both u and x are nonnull and $ux^*v \subseteq L$. Since $uxv = w$ is primitive, so is its conjugate xvu. Since xvu is primitive, x and vu cannot be powers of a common word. By Proposition 6.5 (Section 6.8) the set $Q \cap x^+vu$ is infinite. Since each word in ux^+v is a conjugate of a word in x^+vu, the set $Q \cap ux^+v$ is also infinite and since also $ux^*v \subseteq L$, $Q \cap L$ is infinite as was to be proved.

Suppose now that $|L \cap Q| = \infty$. Let z be a word of *minimal length* in $L \cap Q$. To conclude the proof it is only necessary to show that $|z| \leq B$: Suppose that $|z| > B$. Since $B = 4n$ and $M(L)$ has only n states, z possesses a factorization $z = ux'xvy'yw$ for which: $|ux'x| < 2n$; $|y'yv| < 2n$; $[u] = [ux'] = [ux'x]$; and $[ux'xv] = [ux'xvy'] = [ux'xvy'y]$; and none of the words x', x, y', y, uvw is null. We are concerned with the relative lengths of the four words x', x, y', and y. It is sufficient to treat only the case in which: $|x| \leq |x'|$, $|y| \leq |y'|$, and $|x'| \leq |y'|$. Each of the seven other settings of the inequalities can be treated in an exactly analogous manner. (See Exercises 1 and 2 in this section, below.)

Since $uxvw$ and $uxxvw$ are in L and are shorter than z, neither is in Q. Consequently, neither of their conjugates $xvwu$ and $xxvwu$ is in Q. From Proposition 6.5 it follows that $rt(x) = rt(vwu)$. Since $uxvy'yw$ and $uxxvy'yw$ are in L and are shorter than z, neither is in Q. Consequently, neither of their conjugates $xvy'ywu$ and $xxvy'ywu$ is in Q. From Proposition 6.5 it follows that $rt(x) = rt(vy'ywu)$. Since $ux'vw$ and $ux'x'vw$ are in L and are shorter than z, neither is in Q. Consequently, neither of their conjugates $x'vwu$ and $x'x'vwu$ is in Q. From Proposition 6.5, it follows that $rt(x') = rt(vwu)$. We now have $rt(x') = rt(x) = rt(vy'ywu)$. Consequently the word $(x')(x)(vy'ywu)$ is not primitive, being in fact a power of $rt(x)$. Since $z = ux'xvy'yw$ is a conjugate of $x'xvy'ywu$ it cannot be primitive either. This contradiction confirms that the shortest word in $L \cap Q$ has length $\leq B$. $\qquad\square$

Exercises

(1) Carry out the proof in the final two paragraphs of Theorem 6.1 above using the settings: $|x| \leq |x'|$, $|y| \leq |y'|$, and $|y'| \leq |x'|$.

(2) Carry out the proof in the final two paragraphs of Theorem 6.1 above using the settings: $|x'| \leq |x|$, $|y| \leq |y'|$, and $|y'| \leq |x|$.

(3) Study the proof of Theorem 6.1 above to see if the given proof will hold if you replace $B = 4n$ by $B = 4n - 1$. Can you reduce B any further without some basic additional insight?

Remark 6.1 *The value of B can be reduced a good deal in Theorem 6.1 and in the resulting Counting Procedure using more powerful tools from word combinatorics. This is confirmed for $B = 3n - 3$ in [25] and later for $B = (1/2)(5n - 9)$ by M. Katsura and S. S. Yu. See also [13].*

6.10 Algorithmic sketching of regular languages

The spectrum of any word w in Σ^+ relative to a regular language L can be read from the flag of w. This is merely a matter of noting which of the states in the flag of w is a final state of $M(L)$. Thus words having the same flag have the same spectrum. There are several absolutely fundamental consequences of this fact: (1) The flag partition P' of Σ^+ refines the spectral partition P; (2) since a regular language has only finitely many flags it has only finitely many distinct spectra; and (3) since each spectral class is the union of (a finite number) of flag languages (each of which is regular) *the spectral classes of a regular language are regular.* Thus both $P'(L)$ and $P(L)$ are finite partitions of Σ^+ into regular sets.

Theorem 6.2 *Each regular language has only finitely many sketch parameters and these parameters are algorithmically computable.*

Proof Given a regular language L, construct $M(L)$. Let m be the number of states of $M(L)$. Only finitely many sequences of states $F = \{s_j : 0 \le j < k\}$ with $k \le m$ could *possibly* occur as flags of words in Σ^+. For each such sequence F construct the intersection $I = \cap\{L(s_j, s_{j+1}) : 0 \le j \le k - 1\}$, where each $L(s_j, s_{j+1})$ is the language that consists of all words x in Σ^+ for which $s_j x = s_{j+1}$. If I is *empty* then F is *not* the flag of any word. If I is *not empty* then F *is* the flag of each word w in I and consequently we have $I(F) = I$. At this point we have determined the partition $P'(L)$ of Σ^+ into the flag languages determined by L. Note that each flag F determines the spectrum that is common to each word w in $I(F)$ since $\text{Sp}(w) = \{n \in N : [w^n]$ is a final state of $M(L)\}$.

For each flag F associated with L, determine the spectrum of F and apply the Counting Procedure in Section 6.9 to determine the cardinal number of $\text{Su}(I(F))$. Since distinct flags may have the same spectrum, flag languages that have a common spectrum must be collapsed together. Each spectral class C arises as the union of the flag languages it contains. Thus the spectral partition $P(L)$ arises as the resulting coarsening of $P'(L)$. Each sketch parameter arises from a spectral class C that contains a primitive word q and has the form $(\text{Sp}(q), sum \{|\text{Su}(I(F))| : I(F) \subseteq C\})$. $\qquad\qquad\square$

An Example Computation Let $L = (a \vee b)a^*b^*$. One may verify that $M(L)$ has four states: $[\lambda]$, $[a] = [b] = [aa] = [ba]$, $[ab] = [bb]$, $[aba] = [bba] = [abab] = [baba] = $ "dead." There are two final states: $[a]$ and $[ab]$ and two nonfinal states $[\lambda]$ and "dead." There are six distinct flags: $F(a) : [\lambda]$, $[a] = [aa]$; $F(b) : [\lambda], [b], [bb] = [bbb]$; $F(ab) : [\lambda], [ab], [abab] = $ "dead;" $F(bb) : [\lambda], [bb] = [bbbb]$; $F(ba) : [\lambda], [ba], [baba] = $ "dead;" and $F(aba) : [\lambda], [aba] = $ "dead." The languages of these six flags are: $L(F(a)) = a^+$; $L(F(b)) = b^+$; $L(F(ab)) = a^+b^+ \vee ba^+b^*$; $L(F(bb)) = bb^+$; $L(F(ba)) = ba^+$; $L(F(aba)) = (a \vee b)^*(aba \vee bba)(a \vee b)^*$. We count the primitive words in each flag language: $L(F(a))$ contains one primitive word, namely a; $L(F(b))$ contains one, namely b; $L(F(ab))$ contains an infinite number of primitive words; $L(F(bb))$ contains no primitive words; $L(F(ba))$ contains infinitely many primitive words and so does $L(F(aba))$. The spectra of these flag languages of primitive words are: $\text{Sp}(a) = N$; $\text{Sp}(b) = N$; $\text{Sp}(ab) = \{1\}$; $\text{Sp}(ba) = \{1\}$; and $\text{Sp}(aba) = \emptyset$. The two flag languages, containing a and

b, respectively, have the same spectrum N. Thus the union of these two flag languages, which is $a^+ \vee b^+$, constitutes a spectral class. The two flag languages, containing ab and ba, respectively, have the same spectrum $\{1\}$. Thus the union of these two flag languages, which is $a^+b^+ \vee ba^+b^* \vee ba^+ = a^+b^+ \vee ba^+b^*$, constitutes a second spectral class. Finally, the flag language containing aba, namely $(a \vee b)^*(aba \vee bba)(a \vee b)^*$, constitutes the third spectral class of L. The first spectral class contains exactly two primitive words, namely, a and b. This gives the parameter: $(N, 2)$. The second spectral class contains infinitely many primitive words. This gives the parameter: $(\{1\}, \infty)$. The third spectral class contains infinitely many primitive words which gives the parameter: $(\emptyset, \infty)$.

Using the sketch parameters from the example above we provide a sketch of L: Let $B : Q \to Z$ be any bijection for which: $B(a) = 1$; $B(b) = 2$; B establishes a one-to-one correspondence between the set of primitive words in the second (infinite) spectral class above and the set $\{z \in Z : z \le 0\}$; and B establishes a one-to-one correspondence between the set of primitive words in the third (infinite) spectral class above and the set $\{z \in Z : z \ge 3\}$. In this sketch of L, there is a vertical black stripe two units wide above a and b (i.e., $x = 1$ and $x = 2$). The remainder of the right quadrant is white. The left quadrant is white except for one horizontal black stripe at the level $n = 1$. Although this language is not bounded above, it is almost uniformly bounded above. It is not almost cylindrical, but it is uniformly eventual. From this sketch of L all further sketches of L can be obtained by permuting the columns of the given sketch.

Corollary 6.1 *Sketch equivalence is decidable for each pair of regular languages. Each of the ten language-theoretic properties defined in Section 6.5 is decidable for a regular language.*

Procedures These decisions can be made after computing the sketch parameters of the languages in question. Two languages are sketch equivalent if and only if they have the same set of sketch parameters. The ten decisions concerning a regular language are easily made by an examination of the sketch parameters of the language.

Which of the two partitions $P(L)$ and $P'(L)$ induced by a language L in the free semigroup Σ^+ is more fundamental may not be clear at this time. In this chapter the detailed work has been done at the flag level, $P'(L)$. A previous exposition [23] applied the algorithms given by M. Ito, H. J. Shyr, and S. S. Yu in their paper [25] to construct the sketch parameters of the regular languages. See also [13] for new elegant short proofs providing relevant tools.

Open Ended Exercise   Let Σ be an alphabet and let K be an arbitrary language contained in Σ^+. If the sketch parameters of K are given, to what extent can they be used to decide whether there is a regular language L that has these sketch parameters? Special cases in which K is required to satisfy one or more of the ten language-theoretic properties defined in Section 6.5 might be treated.

Open Ended Exercise   Can additional classes of languages be found that allow their sketches to be determined? Note that for each *context-free* language L, $w^* \cap L$ is *regular* and recall Exercise 2 of Section 6.3.

Open Ended Project   The production of software for displaying sketches of languages is encouraged.

An Aside to Readers Interested in Art   The inspiration for the vision-based approach to languages came in part from admiration for the late paintings of Piet Mondrian and certain paintings by Barnet Newman. Note that one can sketch two or more languages on the same half plane and use distinct color pairs for distinct languages.

7

From biopolymers to formal language theory

7.1   Introduction

Living systems on our planet rely on the construction of long molecules by linking relatively small units into sequences where each pair of adjoining units is connected in a uniform manner. The units of polypeptides (proteins) are a set of twenty amino acids. These units are connected by the carboxyl group (COOH) of one unit being joined through the amino group (NH_2) of the next unit, with a water molecule being deleted in the process. The units of RNA are a set of four ribonucleotides. These units are connected by the phosphate group (PO_4 attached at the $5'$ carbon) of one unit being joined through replacement of the hydroxyl group (OH attached at the $3'$ carbon) of the next unit, with a water molecule being deleted in the process. The units of single stranded DNA are a set of four deoxyribonucleotides with the joining process as in the case of RNA.

Molecules lie in three-dimensional space, whereas words lie on a line. One may adopt the *convention* of listing the amino acids of a protein on a line with the free amino group on the left and the free carboxyl group on the right. For both single stranded RNA and DNA molecules one may adopt the *convention* of listing their units on a line with the phosphate at the left and the free hydroxyl group at the right. These conventions allow us to model (without ambiguity) these biopolymers as words over finite alphabets: a twenty letter alphabet of symbols that denote the twenty amino acids and two four letter alphabets of symbols denoting the four units for RNA and DNA, respectively.

Within a decade of the announcement in 1953 of the structure of DNA by Watson and Crick, mathematicians and scientists were suggesting that bridges be found between the study of the fundamental polymers of life and the mathematical theory of words over abstract alphabets. The biopolymers were modeled by words in a free monoid with word concatenation modeling the chemical end-to-end joining of biopolymers through deletion of water. To obtain nontrivial results in this rarefied context some additional source of structure seems to be required. The Shannon information content of biopolymers has long been studied. The *transcription* of DNA into RNA and the *translation* of RNA into protein are easily viewed as actions of *finite transducers*. This chapter treats additions to formal language theory that have their source in the conceptual modeling of the actions of enzymatic processes on double stranded DNA. The modeling process has motivated new concepts, constructions, and results in the theory of formal languages and automata. The focus of this chapter is on these new concepts, rather than the associated biomolecular science. Discussion of the science that provoked these developments in formal language theory has been restricted to this section and Section 7.3, with Section 7.3 optional reading for the interested reader.

Section 7.2 is an informal introduction to what are called **splicing** operations using examples that may appear quite arbitrary at first reading. Those who read the optional Section 7.3 will find that the examples of Section 7.2 are abstractions of the "cut and paste" actions of commercially available enzymes operating on DNA molecules. Section 7.4 provides the definitions and constructions required for a formal theory of splicing. In the remaining sections the deepest results relating the theory of splicing systems and the class of regular languages are treated.

Although all of the motivating biomolecular examples given here involve double stranded DNA, splicing theory is potentially applicable to polypeptides, RNA, DNA (whether single stranded or double stranded) and any other polymers that may be viewed as strings of related units linked in a uniform manner. (The cytoskeletal filaments in eukaryotic cells provide several such examples.) Moreover, dsDNA frequently occurs, both *in vivo* and *in vitro*, in circular form. Linear and circular dsDNA molecules interact (inter-splice) in nature as illustrated for ciliate genomes in [27] and [6]. Interactions between linear and circular DNA have been discussed in an abstract splicing context in [18] and [33]. A review of *in vitro* solutions of standard combinatorial computations using the cut and paste operations discussed here appears in [20]. The intention of this chapter is to stimulate the creation of additional connections between formal language theory and the biomolecular sciences.

7.2   Constructing new words by splicing together pairs of existing words

Given an *ordered pair* of words over the alphabet $\{a, c, g, t\}$, for example $u = ttttggaaccttt$ and $v = tttggaacctttt$, one can consider allowing the construction of a new word from these two by "*cutting*" each, say between the two occurrences of the symbol a in the subsequence $s = ggaacc$ that occurs in each of u and v, and then building a new string by "*pasting*" together (concatenating) the left portion of the first string and the right portion of the second string. The cutting process applied to the ordered pair u and v gives the fragments: *ttttgga, accttt* and *tttgga, acctttt*. The pasting of the indicated fragments in the indicated order then gives the word $x = ttttggaaccttt$. In this way the ordered pair of words u, v has provided x. Note that the ordered pair v, u following the same cut and paste operation produces $y = tttggaacctttt$. We say that we have **spliced** u, v producing x and we have **spliced** v, u producing y.

An extensive literature has developed in which the generative power of such splicing operations on words has been investigated. Many carefully considered *control structures* have been studied that guide the splicing process. Numerous researchers have been able to demonstrate that, by applying various such control structures, they can provide universal (Turing equivalent) computational power based on splicing operations. We do not pursue this goal here; we stay in the realm of regular languages. The original motivation for the introduction of the splicing concept was the modeling of the cut and paste actions provided by sets of restriction enzymes acting on double stranded DNA (dsDNA) molecules. These enzymatic actions are fundamental tools of genetic engineering. Our goal is to show that the theory of regular languages provides a formalism through which the potential generative power of sets of restriction enzymes acting on dsDNA can be represented. Readers who have interdisciplinary inclinations can continue with studies of [**22**] and models of computation based on biochemistry [**32**]. In Section 7.3 a *minimal* discussion of DNA splicing is given to indicate the contact point between splicing as understood in formal language theory and in molecular biology. A reader who does not wish for additional motivation from the biomolecular sciences may skip Section 7.3 which is not required for an understanding of the discussions in later sections.

7.3 The motivation from molecular biology

A single stranded DNA (ssDNA) molecule can be viewed as a linear sequence of the four covalently bonded deoxyribonucleotides {A = adenine, C = cytosine, G = guanine, T = thymine}. For example:

TTTTGGAACCTTT.

A dsDNA molecule can be viewed as a linear sequence of hydrogen bonded pairs where the hydrogen bonds are between the vertically displayed pairs:

TTTTGGAACCTTT
AAAACCTTGGAAA.

It is adequate here to assume that A and T pair only with each other and C and G pair only with each other. Due to this so-called Watson–Crick pairing rule, when one strand of a dsDNA molecule is determined the other is also known. If (as above) one row is

TTTTGGAACCTTT

we know its companion row is

AAAACCTTGGAAA.

Consequently, we need to give only one of the two strands. For efficiency and convenience we will list only one row of each dsDNA molecule. To be certain not to confuse dsDNA and ssDNA, we will use lowercase a, c, g, t to denote the paired deoxyribonucleotides:

A   C   G   T
T   G   C   A

respectively.

Thus TTTTGGAACCTTT is an ssDNA, but ttttggaaccttt is a dsDNA having as one of its strands the ssDNA, TTTTGGAACCTTT.

There are over 200 commercially available restriction enzymes that cut dsDNA molecules at specific subsequences (sites). The example given in Section 7.2 is, in fact, a representation of an actual enzymatic process. At an occurrence of the site ggaacc in a dsDNA molecule the enzyme Nla IV cuts the covalent bonds in each of the single strands that hold the middle a–a of the site together. When this cut is made in aqueous solution the left and right halves separate due to Brownian motion. The resulting freshly cut ends of the fragments can be again connected with restored covalent bonds if an enzyme called a ligase is present. Suppose now that we have a test tube which contains, dissolved in water (or more precisely, in an appropriate *buffer* solution), the dsDNA molecules u = ttttggaacettt and v = tttggaaccttt and also Nla IV and a ligase. Then Nla IV will cut the two molecules u and v producing the four fragments ttttgga', 'accttt and tttgga', 'acctttt, where the ' symbols have been added to denote the freshly cut ends (technically, the phosphates attached at the 5'-ends remain after the cutting and are required for future pasting). The ligase can now paste together the fragments ttttgga' and 'acctttt to yield the dsDNA molecule x = ttttggaacctttt. The ligase can also paste together the fragments tttgga' and 'accttt to yield the dsDNA molecule $y =$ tttggaaccttt. The molecules x and y are said to be **recombinants** of u and v. For completeness we mention that the ligase also has the potential for reconstructing the original molecules u and v from the four fragments. If we ignore any remaining fragments that have freshly cut ends, then we may say that the "language" of all possible molecules that can arise in our test tube consists of the molecular varieties u, v, x, and y. Some significant details concerning DNA molecules have been suppressed above. The interested reader can find these details treated in the following exercise and the references.

Exercise

The two ends of an ssDNA molecule exhibit distinct structures. At one end a methyl group protrudes which may have an attached phosphate group. This end is referred to as the $5'$ end since the carbon atom of the methyl group is counted as the $5'$ carbon of the sugar substructure to which it belongs. At the other end a hydroxyl is attached at the $3'$ carbon of the sugar substructure to which it belongs. This end is referred to as the $3'$ end of the molecule. In modeling one must either label the ends or adopt a convention that allows the labels to be known otherwise. The ssDNA molecules $5'$-ACTTGC-$3'$ and $3'$-ACTTGC-$5'$ are *not* representations of the same molecule. For dsDNA one must understand that the two strands of the molecule always have *opposite* $5' \to 3'$ orientation. For convenience and concision we use the *convention* illustrated here. When, for example, acttgc is used to represent a dsDNA molecule it must be understood that this molecule has as one strand $5'$-ACTTGC-$3'$ and consequently that the dsDNA molecule when fully spelled out is:

$$5'\text{-ACTTGC-}3'$$
$$3'\text{-TGAACG-}5'$$

(a) Write $3'$-ACTTGC-$5'$ with the $5'$ end on the left (and the $3'$ on the right).

(b) Write a lowercase representation for the dsDNA molecule that has $3'$-ACTTGC-$5'$ as one of its strands. Is there a second lowercase representation? Is there a third?

(c) Which pairs of words, when regarded as models of dsDNA molecules, denote the same molecules: acttgc, cgttca, gcaagt, tgaacg, aaattt, tttaaa.

(d) Verify that each dsDNA molecule, when denoted using the alphabet {a,c,g,t} and the conventions established here, has either exactly two distinct representations or only one representation. Give examples of each type. Those having only one are said to possess **dyadic symmetry**. (That dsDNA molecules may possess two distinct word representations creates only a slight nuisance when constructing splicing models as explained in [17] [21] and [22].)

7.4　Splicing rules, schemes, systems, and languages

The previous sections have been written in an informal manner, possibly allowing ambiguity between molecules and the words used to represent them. The remainder of this chapter deals specifically with words in a free monoid. (However, all results in the chapter have meaningful interpretations for enzymes acting on dsDNA.)

Let Σ be a finite set to be used as an alphabet. Let Σ^* be the set of all strings over Σ. By a **language** we mean a subset of Σ^*. A **splicing rule** is an element $r = (u, u', v', v)$ of the product set

$$[\Sigma^*]^4 = \Sigma^* \times \Sigma^* \times \Sigma^* \times \Sigma^*.$$

The **action** of the rule r on a *language* L defines the language $r(L) = \{xuvy$ in $\Sigma^* : L$ contains strings $xuu'q$ and $pv'vy$ for some $x, q, p,$ and y in $\Sigma^*\}$. For each set, R, of splicing rules we extend the definition of $r(L)$ by defining $R(L) = \cup\{r(L) : r \in R\}$. A rule r **respects** the language L if $r(L)$ is contained in L and a set R of rules **respects** L if $R(L)$ is contained in L. By the **radius** of a splicing rule (u, u', v', v) we mean the maximum of the lengths of the strings u, u', v', v.

Definition 7.1　*A splicing scheme is a pair $\sigma = (\Sigma, R)$, where Σ is a finite alphabet and R is a finite set of splicing rules. For each language L and each nonnegative integer n, we define $\sigma^n(L)$ inductively: $\sigma^0(L) = L$ and, for each nonnegative integer k, $\sigma^{k+1}(L) = \sigma^k(L) \cup R(\sigma^k(L))$. We then define $\sigma^*(L) = \cup\{\sigma^n(L) : n \geq 0\}$. A splicing system is a pair (σ, I), where σ is a splicing scheme and I is a finite initial language contained in Σ^*. The language generated by (σ, I) is $L(\sigma, I) = \sigma^*(I)$. A language L is a splicing language if $L = L(\sigma, I)$ for some splicing system (σ, I).*

Example 7.1　Let $\Sigma = \{a, c, g, t\}$. Let $r = (u, u', v', v)$ where the four words u, u', v', v in Σ^* appearing in the rule r are $u = v' = gga$ and $u' = v = acc$. Let $R = \{r\}$. This gives the splicing scheme

$$\sigma = (\Sigma, R) = (\{a, c, g, t\}, \{(gga, acc, gga, acc)\}).$$

Let

$$I = \{ttttggaaccttt, tttggaacctttt\}.$$

Observe that r applied to the ordered pair

$$(ttttggaaccttt, tttggaaccttttt)$$

of words in I gives the word $ttttggaaccttttt$, and r applied to the ordered pair

$$(tttggaacctttt, ttttggaaccttt)$$

of words in I gives $tttggaaccttt$. The less interesting actions of r on I must be recognized: When r acts on ordered pairs in the "diagonal" of $I \times I$, for example on

$$(ttttggaaccttt, ttttggaaccttt)$$

the result is merely $ttttggaaccttt$ which appeared as each coordinate of the pair. Here we have

$$\sigma^0(I) = I = \{ttttggaaccttt, tttggaccttttt\}$$

and

$$\sigma^1(I) = \sigma^0(I) \cup R(\sigma^0(I))$$
$$= I \cup \{ttttggaacctttt, tttggaaccttt, ttttggaaccttt, tttggaacctttt\}$$

equals

$$\{ttttggaacctttt, tttggaaccttt, ttttggaaccttt, tttggaccttttt\}.$$

Notice that R respects $\sigma^1(I)$ and consequently $\sigma^2(I) = \sigma^1(I)$. Then also $\sigma^3(I)) = \sigma^2(I) = \sigma^1(I)$ and in fact $\sigma^*(I) = \sigma^1(I)$. Thus $L(\sigma, I)$ is the finite language

$$\sigma^*(I) = \{ttttggaacctttt, tttggaaccttt, ttttggaaccttt, tttggaacctttt\}.$$

This example connects the formal definitions of splicing systems and languages with the less formal introductory remarks of Sections 7.1 and 7.2.

Example 7.2 Let $\Sigma = \{a, c, g, t\}$. Let $r = (c, cccgg, c, cccgg)$, $R = \{r\}$, and let I contain only one word of length 30,

$$I = \{aaaaaaccccggaaaaaaccccggaaaaaa\}.$$

The rule can be applied to the ordered pair

$$(a^6ccccgga^6ccccgga^6, a^6ccccgga^6ccccgga^6)$$

with cuts made using the right occurrence of $ccccgg$ in the first coordinate and the left occurrence of $ccccgg$ in the second coordinate. This gives the word of length 42:

$$a^6ccccgga^6ccccgga^6ccccgga^6.$$

The rule can be also applied to the ordered pair using the left occurrence of $ccccgg$ in the first coordinate and the right occurrence of $ccccgg$ in the second coordinate. This gives the word of length 18, $a^6ccccgga^6$. Thus

$$\sigma^1(I) = \{a^6ccccgga^6, a^6ccccgga^6ccccgga^6, a^6ccccgga^6ccccgga^6ccccgga^6\}.$$

Continuing with similar considerations one finds that $L(\sigma, I) = \sigma^*(I)$ is the infinite regular language

$$a^6ccccgga^6(ccccgga^6)^*.$$

Example 7.3   We may interpret the 30 symbol word given in Example 7.2 as a model of a dsDNA molecule as indicated in Section 7.3. The rule r of Example 7.2 represents the cut and paste activity of the restriction enzyme BsaJ I accompanied by a ligase. With these understandings the language

$$L(\sigma, I) = a^6ccccgga^6(ccccgga^6)^*$$

obtained in Example 7.2 is a model of the set of all dsDNA molecules (having no freshly cut ends) that can potentially arise in a test tube containing BsaJ I, a ligase, and (sufficiently many) dsDNA molecules of model $a^6ccccgga^6ccccgga^6$. The ability to make assertions as in the preceding sentence motivated the introduction of the splicing concept into formal language theory.

Example 7.4   Let $\Sigma = \{a, b, c\}$. Let L be the regular language $caba^*b$. Can we find a splicing system that generates L? Yes, this can be done very easily by taking advantage of the fact that the symbol c occurs as the leftmost symbol of every word in L and occurs nowhere else in any word of L. Let $r = (caba, a, cab, a)$ and let $I = \{cabb, cabab, cabaab\}$. Note that r allows the generation of $cabaaab$ as follows: From the ordered pair $(cabaab, cabaab)$, and the two distinct conceptual analyses $caba/ab$ and cab/aab, the rule r gives the new word $cabaaab$. Then from the ordered pair $(cabaab, cabaaab)$, and the analyses $caba/ab$ and $cab/aaab$, the rule r gives the word $cabaaaab$. (Note that r provides a form of pumping.) Continuing in this way all words $caba^nb$ with $n \geq 3$ can be obtained. Since $caba^nb$ with $0 \leq n \leq 2$ were given in I, we have, for $R = \{r\}$ and $\sigma = (\Sigma, R)$, $L(\sigma, I) = caba^*b$ as desired. In fact it has been shown [12] that for any regular language L' over any alphabet Σ, by choosing a symbol not in Σ, say c, the language

$$L = cL' = \{cw : w \in L'\}$$

is generated by a splicing system that can be specified very much as we have done for the language $caba^*b$ in this example. Thus informally speaking, each regular language is almost a splicing language.

Example 7.5 Let $\Sigma = \{a, b\}$. The regular language $L = (aa)^*$ cannot be generated by a splicing system. As the reader may verify, any finite set of rules that allows every word in L to be generated will also generate strings of odd length as well as the strings of even length.

Example 7.6 The regular language $L' = a^*ba^*ba^*$ cannot be generated by a finite set of rules either: For any nonnegative integer n,

$$R_n = \{(\lambda, ba^n b, \lambda, aba^n b), (ba^n ba, \lambda, ba^n b, \lambda)\}$$

and

$$I_n = \{aba^n b, ba^n b, ba^n ba\}$$

generate $a^*ba^n ba^*$.

Consequently, for any finite subset F of nonnegative integers, $R = \cup\{R_n : n \in F\}$ and $I = \cup\{I_n : n \in F\}$ generate the language $L'' = \cup\{a^*ba^n ba^* : n \in F\}$. However, as the reader may verify, any finite set of rules and finite initial language that generate all words in $a^*ba^*ba^*$ will also generate words in which the symbol b occurs more than twice. Thus there are regular languages that are not splicing languages.

Exercises

(1) Let L be any finite language over any alphabet Σ. Specify a splicing system that generates L. (Hint: The set R of rules can be empty.)

(2) Let $\Sigma = \{a, b\}$. Find three splicing systems that generate, respectively, (i) $L = b(aa)^*$; (ii) $L = \Sigma^*$; and (iii) $L = ba^*ba^*$.

(3) Let $\Sigma = \{a, b, c\}$. Find three splicing systems that generate, respectively, (i) $L = ab^*abc$; (ii) $L = ab^*cab$; (iii) $a^*ba^*ca^*ba^*$.

7.5 Every splicing language is a regular language

Splicing languages were introduced in published form for the first time in 1987 [17]. Fortunately K. Culik II and T. Harju quickly announced in 1989 [4], [5] that all splicing languages are regular. A second exposition of the regularity result was given by D. Pixton in 1996. This exposition provided, for each splicing system, an explicit construction of a finite automaton that was concisely proved, using an insightfully constructed inductive set, to recognize the language generated by the splicing system. In 1989, R. Gatterdam observed [11] that not all regular languages are splicing languages. So, which regular languages are splicing languages? We would like to have a theorem that characterizes the class of splicing languages in terms of previously known classes of languages. As yet we have no such characterization. In [17] it was observed that a language L is a splicing language if there is a positive integer n such that uxq is always in L whenever x has length n and both uxv and pxq are in L. These languages, which were analyzed rather thoroughly in [19], constitute a highly restricted subclass of the splicing languages, enlargements of which have been studied extensively in [12]. The interested reader is urged to study very carefully Pixton's proof of regularity which is given in [33], [21], [32], [22] and broadly generalized in [34].

With no crisp characterization of the class of splicing languages found, concern turned to the search for an algorithm for deciding whether a given regular language can be generated by a splicing system. There is, of course, an easily described procedure that is guaranteed to discover that a regular language $L \subseteq \Sigma^*$ is a splicing language if L is a splicing language: For each positive integer n, for each set R of rules of radius $\leq n$, and for each subset I of L consisting of strings of length $\leq n$, decide whether $L(\sigma, I) = L$, where $\sigma = (\Sigma, R)$. Since both L and each such $L(\sigma, I)$ are regular, all these steps can be carried out. The procedure terminates when a system $L(\sigma, I)$ is found, but fails to terminate when L is not a splicing language. From this triviality, however, it follows that an algorithm will become available immediately if, for each regular language L, a bound, $N(L)$, can be calculated for which it can be asserted that L cannot be a splicing language unless there is a splicing system having rules of radius $\leq N(L)$ and initial strings of length $\leq N(L)$. (Recall from Section 7.2 that the radius of a rule $r = (u, u', v', v)$ is the length of the longest of the four words u, u', v', and v.) The determination of the bound $N(L)$ is a conceptual victory for the concept of the syntactic monoid of a language because it allows the concise statement of an adequate bound $N(L)$, given in Section 7.7. It also provides a valuable tool stated in the heading of Section 7.6.

Exercises

(1) Let $\Sigma = \{a, b\}$. Find a regular expression that represents $L(\sigma, I)$ where $\sigma = (\Sigma, R)$, $R = \{r\}$, $r = (b, b, a, b)$, and $I = \{abba\}$.

(2) Let $\Sigma = \{a, b, c\}$. Find a regular expression that represents $L(\sigma, I)$ where $\sigma = (\Sigma, R)$, $R = \{r_1, r_2\}$, $r_1 = (ab, c, cb, a)$, $r_2 = (cb, a, ab, c)$, and $I = \{abc, cba\}$.

(3) Same as Exercise 2 with one new rule added: $r_3 = (cbc, \lambda, cb, c)$ so that $R = \{r_1, r_2, r_3\}$.

7.6 The syntactic monoid of a regular language L allows an effective determination of the set of all splicing rules that respect L

First we show how to decide whether a given splicing rule r respects a given regular language $L \subseteq \Sigma^*$: Let M be the minimal automaton recognizing L. Let S be the set of all states of M and let F be the set of final states. For each state s in S and each word w in Σ^* we denote by sw the state of M arrived at after w is read from state s. Note that the rule $r = (u, u', v', v)$ respects L if and only if, for each ordered pair of states p, q of M, for which $\{x \in \Sigma^* : puu'x \in F\}$ and $\{y \in \Sigma^* : qv'vy \in F\}$ are not empty, $\{z \in \Sigma^* : qv'vz \in F\} \subseteq \{z : puvz \in F\}$. The emptiness conditions and the inclusion are decidable since each of the four sets is regular.

Next we show how to specify all of the rules that respect the regular language L in a manner that requires that the procedure above be used on only a finite number of rules. Recall that the syntactic congruence relation, C, in Σ^* is defined by setting $u'Cu$ if and only if, for every pair of strings x and $y \in \Sigma^*$, either both $xu'y$ and xuy are in L or neither is in L. Since L is regular, the number of $C-$congruence classes is a positive integer which we denote $n(L)$. Then there are precisely $[n(L)]^4$ ordered quadruples of congruence classes. Let (W, X, Y, Z) be an ordered quadruple of congruence classes. Let $r = (w, x, y, z)$ and $r' = (w', x', y', z')$ be two rules in $W \times X \times Y \times Z$. We verify that r respects L if and only if r' respects L. By the symmetry of the roles of r and r' in the hypothesis, we need only assume that r respects L and verify that then r' must respect L. Suppose that r respects L and that the pair $uw'x'v, sy'z't$ is in L. We need only show that $uw'z't$ is in L: From $w'Cw$ we have $uwx'v$ is in L and from $x'Cx$ we then have $uwxv$ in L. From $y'Cy$ and $z'Cz$ it follows that $syzt$ is in L. Since r respects L and the pair $uwxv, syzt$ is in L, we have $uwzt$ in L. From $w'Cw$ and $z'Cz$ it follows that $uw'z't$ is in L, as required. Thus, to specify all the rules that respect L, we construct the $[n(L)]^4$ quadruples of syntactic classes determined in Σ^* by L and, from each such quadruple (W, X, Y, Z), we choose one word from each class to obtain one rule (w, x, y, z) and then decide whether it respects L. If it does then every rule in $W \times X \times Y \times Z$ respects L. If it does not respect L then no rule in $W \times X \times Y \times Z$ respects L. This discussion has justified the following:

Proposition 7.1 *Let L be a regular language. The set of rules that respect L has the form*

$$\cup\{W_i \times X_i \times Y_i \times Z_i : 1 \leq i \leq m\}$$

where m is a nonnegative integer and each of the sets

$$W_i, X_i, Y_i, Z_i (1 \leq i \leq m)$$

is an element of the syntactic monoid of L.

Since each syntactic class of a regular language L is itself a regular language, one can list all the strings of length at most k in the class. Consequently when the representation in the proposition has been constructed, the set of all rules of radius at most k that preserve L can be listed with no additional testing: For each of the sets

$$W_i \times X_i \times Y_i \times Z_i (1 \leq i \leq m)$$

in the representation, list all of the rules (w, x, y, z) in

$$W_i \times X_i \times Y_i \times Z_i$$

of radius at most k. In order to create such a list without using the syntactic monoid it would be necessary to list every rule of radius at most k in all of $[\Sigma^*]^4$ and test every such rule individually to decide if it preserves L.

Exercises

(1) Let $\Sigma = \{a\}$ and $L = (aa)^*$. Construct the syntactic monoid of L.
(2) Let $\Sigma = \{a, b\}$ and $L = a^*ba^*ba^*$. Construct the syntactic monoid of L.
(3) Construct the syntactic monoid of the language in Exercise 3 of Section 7.5.

7.7   It is algorithmically decidable whether a given regular language is a reflexive splicing language

A rule set R is **reflexive** if, for each rule (u, u', v', v) in R, the rules (u, u', u, u') and (v', v, v', v) are also in R. When R is reflexive we say the same of any scheme or system having R as its rule set. In fact, splicing systems that model the cut and paste action of restriction enzymes and a ligase are necessarily reflexive. Consequently, from a modeling perspective, it is the reflexive splicing systems that are of prime interest.

Section 7.6 provides the tools to construct, for each regular language L and each positive integer k, the following finite reflexive set T_k of splicing rules:

$$T_k = \{(u, u', v', v) : \text{ the radius of } (u, u', v', v) \leq k \text{ and each of the three rules}$$
$$(u, u', v', v), (u, u', u, u'), \text{ and } (v', v, v', v) \text{ respects } L\}.$$

Recall that $T_k(L) = \cup\{r(L) : r \in T_k\}$, which is regular since T_k is finite and, since L is regular, each $r(L)$ is regular (as confirmed in Exercise 5 of this section). Consequently $L \setminus T_k(L)$ is also regular.

Theorem 7.1 (Pixton and Goode)   *A regular language L is a reflexive splicing language if and only if $L \setminus T_k(L)$ is finite where $k = 2(n(L)^2 + 1)$ and $n(L)$ is the cardinal number of the syntactic monoid of L.*

Let L be a regular language. In Chapter 3 the syntactic monoid of L was defined in a way that allows $n(L)$, and therefore also k, to be computed. Section 7.6 provides a procedure for computing the finite set T_k, from which the regular set $T_k(L)$ can be computed and it can be decided whether $L \setminus T_k(L)$ is finite. Thus the theorem of Pixton and Goode provides an algorithm that allows one to decide whether any given regular language is a reflexive splicing language. It is tempting to suppose that when $L \setminus T_k(L)$ is finite it can serve as the set of initial words of a splicing system that generates L. Unfortunately this is not the case as shown in Exercise 3 of this section. Although the proof of the theorem is beyond the scope of this book, the decision procedure that it provides can, in principle, be carried out using the machinery this book has provided.

Exercises

(1) Let $\Sigma = \{a\}$ and $L = (aa)^*$. Compute $n(L)$ and k for this language.
(2) Let $\Sigma = \{a, b\}$ and $L = a^*ba^*ba^*$. Compute $n(L)$ and k for this language.
(3) Let $\Sigma = \{a, b\}$ and $L = \Sigma^*$. Note that uCv for every u, v in L and consequently the syntactic monoid of L is a singleton.
   (a) Compute $n(L)$ and k.
   (b) Describe T_k in words. How many elements does T_k contain?
   (c) Compute $T_k(L)$ and $L \setminus T_k(L)$.
   (d) Conclude that, when the set $L \setminus T_k(L)$ is finite, it does not follow that $L \setminus T_k(L)$ is adequate to serve as the set I of initial words of any splicing system (σ, I) for which $L = L(\sigma, I)$. (This is what makes the proof of Theorem 7.1 challenging.)
   (e) Without using Theorem 7.1, specify a set R of splicing rules and a set I of initial strings for which $L = L(\sigma, I)$ for $\sigma = (\Sigma, R)$. Hint: $(\lambda, \lambda, \lambda, \lambda)$ is a splicing rule.

(4) Show that the following definition of a reflexive splicing language is equivalent to the definition given: A language L is a **reflexive** splicing language if $L = L(\sigma, I)$ for some splicing scheme $\sigma = (\Sigma, R)$ and, for each rule $r = (u, u', v', v)$ in R, the rules (u, u', u, u') and (v', v, v', v) respect L. (This alternative definition allows one to list fewer rules in the rule set specifying a reflexive splicing system.)

(5) Let $L \subseteq \Sigma^*$ be a regular language. Let $r = (u, u', v', v)$ be a splicing rule with u, u', v', v in Σ^*. Show that $r(L)$ is regular. Hint: Use two copies, M and M', of the minimal automaton M that recognizes L. Let the sets of states of these two automata be S and S'. Combine M and M' into a single automaton M'', having state set $S \cup S'$, by adding carefully chosen new edges that allow transitions from states in S to states in S' having v as label. Choose the initial state i of M as the initial state of M'' and choose the set F' of final states of M' as the set of final states of M''.

Appendix A
Cardinality

Theorem A.1 *If $|S| \leq |T|$ and $|T| \leq |S|$, then there is a one-to-one correspondence between S and T, i.e. $|S| = |T|$*

Proof Assume $f : S \to T$ and $g : T \to S$ are injective functions. For each $s \in S$, we find $g^{-1}(s)$ if it exists. We then find $f^{-1}g^{-1}(s)$ if it exists. Then find $g^{-1}f^{-1}g^{-1}(s)$ if it exists. We continue this process. There are three possible results: (1) The process continues indefinitely. (2) The process ends because for some s_i in the process, there is no $g^{-1}(s_i)$. (3) The process ends because for some t_i in the process, there is no $f^{-1}(t_i)$. Let S_1 be the elements of S for which the first result occurs. Let S_2 be the elements of S for which the second result occurs. Let S_3 be the elements of S for which the third result occurs. Obviously these sets are disjoint. Similarly form T_1, T_2, and T_3 as subsets of T. f is a one-to-one correspondence from S_1 to T_1. f is also a one-to-one correspondence from S_2 to T_2. g^{-1} is a one-to-one correspondence from S_3 to T_3. Let $\theta : S \to T$ be defined by

$$\theta(s) = f(s) \text{ if } s \in S_1$$
$$= f(s) \text{ if } s \in S_2$$
$$= g^{-1}(s) \text{ if } s \in S_3$$

θ is a one to one correspondence from S to T. □

Theorem A.2 *For any set A, $|A| < |\mathcal{P}(A)|$.*

Proof Certainly $|A| \leq |\mathcal{P}(A)|$ since for each element in a in A, $\{a\}$ is in $\mathcal{P}(A)$. Assume $|A| = |\mathcal{P}(A)|$. Then there is a one-to-one correspondence between $|A|$ and $|\mathcal{P}(A)|$. For $a \in A$ let $\phi(a)$ be the element in $\mathcal{P}(A)$ paired with a. Some elements in A belong to the element in $\mathcal{P}(A)$ with which they are paired. For example, if $a \in A$ and $\phi(a) = A$ in $\mathcal{P}(A)$, then $a \in \phi(a)$. However, if $a \in A$ and $\phi(a) = \emptyset$, the empty set in $\mathcal{P}(A)$, certainly $a \notin \phi(a)$. Let

245

$W = \{a : a \in A \; a \notin \phi(a)\}$. $W \in \mathcal{P}(A)$), but no element in A can correspond to W, for if $\phi(a) = W$ and $a \in W$, then by definition of W, $a \notin \phi(a)$ and $a \notin W$. However, if $\phi(a) = W$ and $a \notin W$, then $a \notin \phi(a)$ and by definition of W, $a \in W$. Hence we have a contradiction if any element of A corresponds to W and there is no one-to-one correspondence between $|A|$ and $|\mathcal{P}(A)|$. $\square$

This theorem shows us that, for any infinite set, there is another infinite set with greater cardinality. We shall not prove it here but the cardinality of the real numbers is equal to the cardinality of the power set of the set of integers.

Appendix B

Co-compactness Lemma

Lemma B.1 *(Co-compactness Lemma) Let A be a finite set and let $\{R_i : i \in I\}$ be a family of retracts in A^*. There is a finite subset F of I for which $\bigcap_{i \in F} R_i = \bigcap_{i \in I} R_i$.*

Proof We consider only the case for which there is a single key set K for which, for each $i \in I$, K is a set of keys for the key code that generates R_i. The general result then follows from the fact that there are only a finite number of subsets, hence of possible key sets of A. First we partition K into disjoint subsets K' and K''. Let $K' = \{a \in K$: for every finite subset J of I, a occurs in at least one word in $\bigcap_{i \in J} R_i$. Let $K'' = K - K'$.

From the definition of K'', it follows that, for each $a \in K''$, there is a finite subset $F(a)$ of I for which $\bigcap_{i \in F(a)} R_i$ contains no word in which a occurs. Let $F'' = \bigcup_{a \in K''} F(a)$ a in K''. The symbols in K that occur in words in $\bigcap_{i \in F''} R_i$ are precisely the symbols in K''.

Define an equivalence relation $\sim$ in the set $\bigcap_{i \in I} R_i$: by $R_i \sim R_j$ if for each $a \in K'$, the generator of R_i in which a occurs is identical with the generator of R_j in which a occurs. In the next three paragraphs we show that there are only finitely many $\sim$ equivalence classes.

Choose an arbitrary index $m \in I$. Let C be the key code that generates R_m. Let C' be the subset of C consisting of those words with keys in K'. Let L be the length of the longest word in C'. Note that, for any word $w \in C^*$: (1) the number of symbols to the left of the first occurrence of a key symbol in w is less than or equal to $L - 1$; (2) the number of symbols occurring between two successive occurrences of keys is less than or equal to $2L - 2$; and (3) the number of symbols to the right of the last occurrence of a key symbol in w is less than or equal to $L - 1$.

Next we establish that, for every $j \in I$ and every $k \in K'$, the generator of R_j in which k occurs has length at most $4L - 2$. For such j and k we have: since $G = F'' \cup \{m, j\}$ is finite and $k \in K'$, there is a word $w \in \bigcap_{i \in G} R_i$ in which k occurs. Let $w = x_0 a_0 x_1 a_1 \ldots x_{i-1} a_i x_i a_{i+1} x_{i+1} \ldots x_{n-1} a_n x_n$ where the a_i, $1 \le i \le n$, are all the key occurrences in w. Hence, no key occurs in any of the x_i, $1 \le i \le n$. Note that all of the keys occurring in w must lie in K'. The word w can be segmented into code words belonging to R_m and it can also be segmented into code words belonging to R_j. We have $k = a_i$ for some i, $1 \le i \le n$. Note that, if the length of the code word belonging to R_j in which k occurs were greater than or equal to $4L - 2$, this would contradict one of (1),(2), or (3) of the final sentence of the previous paragraph.

We have shown that there is a bound $B(= 4L - 2)$ such that, for every $j \in I$ and $k \in K'$, no code word of R_j in which k occurs can have length greater than or equal to B. From the definition of the equivalence relation $\sim$ we see that there are only finitely many $\sim$ equivalence classes.

Let F' be a subset of I for which, for each $i \in I$, there is a unique $j \in F'$ for which $R_i \sim R_j$.

The statement of the lemma is true for $F = F' \cup F''$.   □

References

[1] J. A. Anderson, *Semiretracts of a free monoid, Theor. Comput. Sci.* **134**(1994), 3–11.

[2] J. A. Anderson, *"Semiretracts of a free monoid"* (extended abstract) *Proceedings of the Colloquium of Words, Languages and Combinatorics II*, New York, World Scientific (1994), 1–5.

[3] J. A. Anderson and T. Head, *The lattice of semiretracts of a free monoid, Intern. J. Computer Math.* **43**(1992), 127–31.

[4] K. Culik II and T. Harju, The regularity of splicing systems and DNA, in *Proceedings ICALP '89,* Lecture Notes in Comput. Sci. **372**(1989), 222–33.

[5] K. Culik II and T. Harju, Splicing semigroups of dominoes and DNA, *Discrete Appl. Math.* **31**(1991), 261–77.

[6] A. Ehrenfeucht, T. Harju, I. Petre, D. M. Prescott and G. Rozenberg, *Computation in Living Cells – Gene Assembly in Ciliates*, Berlin, Springer-Verlag (2004).

[7] J. Engelfriet and G. Rozenberg, *Equality languages and fixed point languages, Information and Control* **43**(1979), 20–49; **31**(1991), 261–77.

[8] L. Fuchs, *Infinite Abelian Groups*, New York, Academic Press, **I** (1970), **III** (1973).

[9] W. Forys and T. Head, *Retracts of free monoids are nowhere dense with respect to finite group topologies and p-adic topologies, Semigroup Forum,* **42**(1991), 117–19.

[10] W. Forys and T. Head, *The poset of retracts of a free monoid, Intern. J. Computer Math.* **37**(1990), 45–8.

[11] R. W. Gatterdam, Splicing systems and regularity, *Intern. J. Computer Math.* **31**(1989), 63–7.

[12] E. Goode (Laun), *Constants and Splicing Systems*, Dissertation Binghamton University, Binghamton, New York (1999).

[13] T. Harju and D. Nowotka, The equation $x^i = y^j z^k$ in a free semigroup, *Semigroup Forum* **68**(2004), 488–90.

[14] J. Harrison, On almost cylindrical languages and the decidability of the D0L and PWD0L primitivity problems, *Theor. Comput. Sci.* **164**(1996), 29–40.

[15] T. Head, Cylindrical and eventual languages, in *Mathematical Linguistics and Related Topics* (Gh. Paun, ed.), Bucharest, Romania, Editura Academiei Romane (1995), pages 179–83.

[16] T. Head, *Expanded subalphabets in the theories of languages and semigroups, Intern. J. Computer Math.* **12**(1982), 113–23.

249

[17] T. Head, Formal language theory and DNA: an analysis of the generative capacity of specific recombinant behaviors, *Bull. Math. Biology* **49**(1987), 737–59.

[18] T. Head, Splicing schemes and DNA, in *Lindenmayer Systems – Impacts on Theoretical Computer Science, Computer Graphics, and Developmental Biology* (G. Rozenberg and A. Salomaa, eds.), Berlin, Springer-Verlag (1992), pages 371–83.

[19] T. Head, Splicing representations of strictly locally testable languages, *Discrete Appl. Math.* **87**(1998), 139–47.

[20] T. Head and S. Gal, Aqueous computing: writing on molecules dissolved in water, in *Nanotechnology: Science and Computation* (J. Chen, N. Jonoska and G. Rozenberg, eds.), Berlin, Springer-Verlag (2006), pages 321–34.

[21] T. Head, Gh. Paun and D. Pixton, Language theory and molecular genetics: generative mechanisms suggested by DNA recombination, in *Handbook of Formal Languages* (G. Rozenberg and Gh. Paun, eds.), Berlin, Springer-Verlag (1997), Vol. II, Chapter 7.

[22] T. Head and D. Pixton, Splicing and regularity, in *Formal Languages and Applications II* (Z. Esik, C. Martin-Vide and V. Mitrana, eds.), Berlin, Springer-Verlag (2006).

[23] T. Head, Visualizing languages using primitive powers, in *Words, Semigroups, & Transducers* (M. Ito, Gh. Paun, and S. Yu, eds.), Singapore, World Scientific (2001), pages 169–80.

[24] T. Head and B. Lando, Periodic D0L languages, *Theor. Comput. Sci.* **46**(1986), 83–9.

[25] M. Ito, M. Katsura, H. J. Shyr and S. S. Yu, Automata accepting primitive words, *Semigroup Forum* **37**(1988), 45–52.

[26] B. Lando, Periodicity and ultimate periodicity of D0L systems, *Theor. Comput. Sci.* **82**(1991), 19–33.

[27] L. F. Landweber and L. Kari, Universal molecular computation in ciliates, in *Evolution as Computation* (L. F. Landweber and E. Winfree, eds.), Berlin, Springer-Verlag (2002), pages 3–15.

[28] M. Lothaire, *Combinatorics on Words*, Massachusetts, Addison-Wesley (1983). Reprinted by Cambridge University Press (1984).

[29] G. H. Mealy, A Method for Synthesizing Sequential Circuits, *Bell System Technical Journal* **34**(1955), 1045–70.

[30] E. F. Moore, Gedanken-Experiments on Sequential Machines, *Automata Studies, Annals of Mathematical Studies* **32**(1956), 129–53.

[31] R. McNaughton and S. Papert, *Counter-Free Automata*, Cambridge, MA, MIT Press (1971).

[32] Gh. Paun, G. Rozenberg and A. Salomaa, *DNA Computing – New Computing Paradigms*, Berlin, Springer-Verlag (1998).

[33] D. Pixton, Regularity of splicing languages, *Discrete Appl. Math.* **69**(1996), 101–24.

[34] D. Pixton, Splicing in abstract families of languages, *Theor. Comput. Sci.* **234**(2000), 135–66.

[35] A. Salomaa, *Jewels of Formal Language Theory*, Rockville, MD, Computer Science Press (1986).

[36] H. J. Shyr, *Free Monoids and Languages*, Taiwan, Hon Min 1991.

[37] B. Tilson, *The intersection of free submonoids of a free monoid is free*, *Semigroup Forum* **4**(1972), 345–50.

Further reading

[1] J. A. Anderson, *Code properties of minimal generating sets of retracts and semiretracts, SEA Bull. Math.* **18**(1994), 7–16.

[2] J. A. Anderson, *"Semiretracts and their syntactic monoids,"* Semigroups: International Conference on Semigroups and its Related Topics, Singapore, Springer-Verlag (1998).

[3] J. Berstel and D. Perrin, *Theory of Codes*, New York, Academic Press (1985).

[4] E. Goode and D. Pixton, Recognizing splicing languages: syntactic monoids and simultaneous pumping, *Discrete Appl. Math.* (to appear).

[5] J. Howie, *Automata and Languages*, New York, Oxford University Press (1991).

[6] J. E. Pin, *Varieties of Formal Languages*, New York, Plenum (1986).

[7] G. Thierrin, *A mode of decomposition of regular languages, Semigroup Forum* **10**(1975), 32–8.

Index

Printed in the United States
By Bookmasters